TACOMA'S TALL SHIP

TACOMA'S TALL SHIP

The Extraordinary Journey of the *Odyssey*

Emily Elizabeth Molina

Published by The History Press
Charleston, SC
www.historypress.com

Cover image: *Odyssey* on Puget Sound with full sails and Ship 190 Sea Scouts out of Tacoma, taken in the early 1990s/mid-2000s. *Courtesy of SSSOdyssey.org.*

First published 2024

Manufactured in the United States

ISBN 9781467157773

Library of Congress Control Number: 2024938213

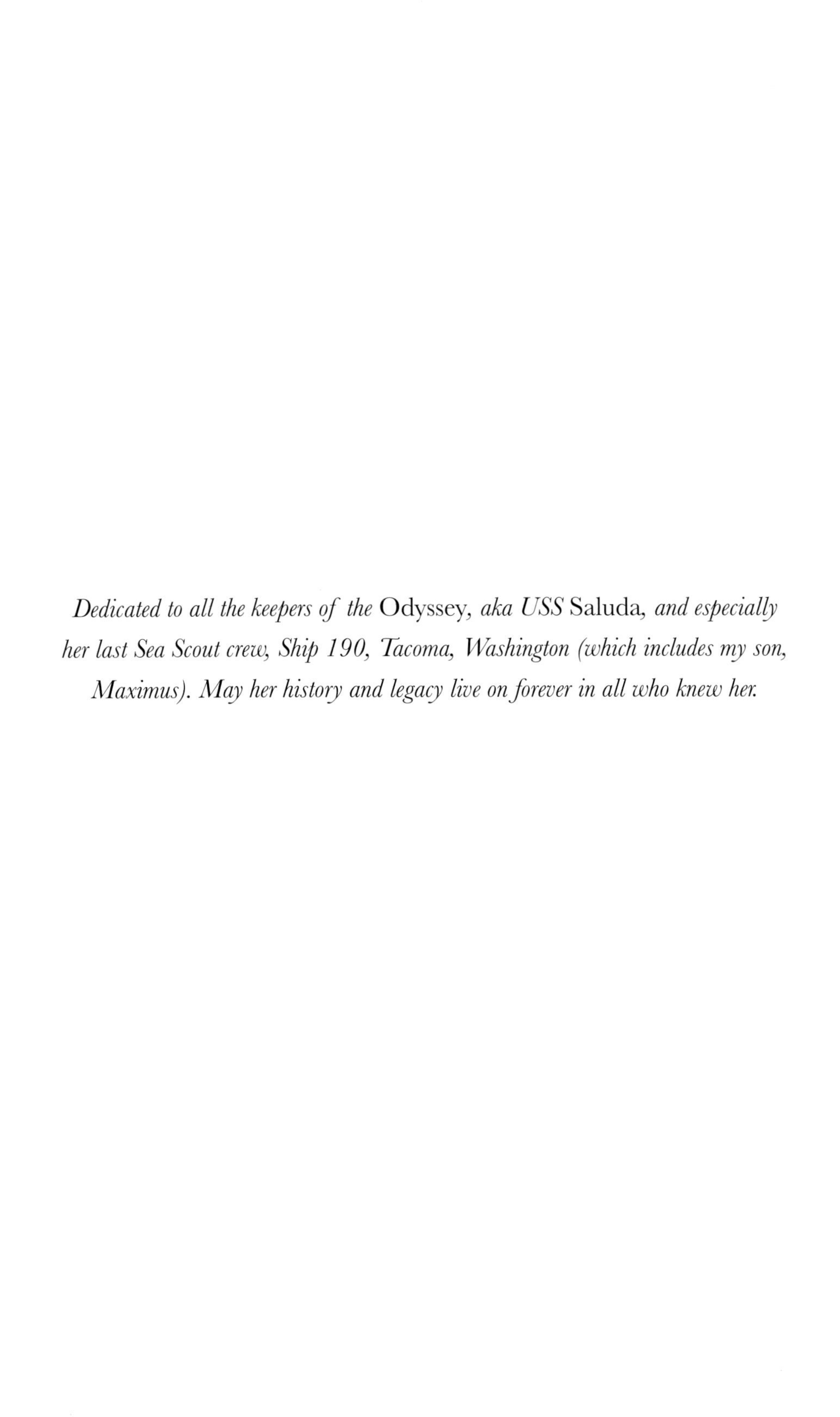

Dedicated to all the keepers of the Odyssey, *aka USS* Saluda, *and especially her last Sea Scout crew, Ship 190, Tacoma, Washington (which includes my son, Maximus). May her history and legacy live on forever in all who knew her.*

CONTENTS

ACKNOWLEDGEMENTS

To my friend Terry Eley and her wonderful story ideas. Without them I would never have stumbled on the *Odyssey* or known what Sea Scouts were.

To Bud Bronson for being there the very first day I laid eyes on *Odyssey*, for giving me my first tour on board and knowing just what to say to get me hooked—and for all of the wonderful Naval Academy anecdotes.

My deepest gratitude to Michiel Hoogstede, an amazing historian, captain and volunteer. His foresight in conducting interviews with Don Frothingham and William "Bill" Gallagher from USS *Saluda* saved material that would otherwise have been lost.

To a chance encounter with Van Chesnutt due to his interest in the *Odyssey* for fictional stories based on USS *Saluda* that led me to his good friend Randy Wall.

Thank you to Randy Wall for sharing personal stories and rare photos passed on by his father, George H. Wall, of him and his school friends who served together on board USS *Saluda* during World War II.

A special thanks to Renée and Terry Paine for keeping and sharing wonderful records, photographs and historical documents of *Odyssey* and for many years dedicated to the Sea Scout program. To Renée's lessons on saloon versus salon—"It's called a saloon. Salons are not on boats. That's where you get your hair done."

Thank you to Rory McDonald for countless years as both a captain and volunteer; for sharing of historical documents, photos and newspapers; and for connecting me with Bob Burns.

To Bob Burns and a wonderful friendship formed out of many *Odyssey* talks; for sharing personal letters written by William Barklie Henry, photographs and historical accounts; and for being an *Odyssey* skipper and captain with the sharpest memory of any person I have ever known, even at ninety-five.

To Linda Barnwell, daughter of William Barklie Henry, for details and information regarding family history and her support of this project.

To Dr. Eric Kiesel, whose support of SSS 190 as a captain, committee chair and volunteer is unparalleled; for teaching me how to sail; and for sharing historical photos and documents passed on to him for safe keeping.

To Don Kidder, former lieutenant commander of the Naval Undersea Center in Long Beach, California, for his willingness to share how much an old wooden boat meant to him through stories about *Saluda*'s time in San Diego.

To Kent Gibson for recollections as the only Sea Scout aboard *Odyssey*'s transit to Tacoma back in 1978, as well as rare photographs from the period.

Thank you to all of the keepers of the *Odyssey*, and a special thanks goes to the last Sea Scout Ship *Odyssey* (Ship 190) youth crew. Your dedication and perseverance are a shining example to all.

INTRODUCTION

When you're a part of something like this, it stays with you forever.
—Renée Paine, early Odyssey *volunteer and Sea Scout supporter*

It was a cold and rainy day in March, quite typical for the Pacific Northwest. There were two wooden boats at the dock that day, and I was very clearly drawn to the old wooden sailboat called *Odyssey*. It was as if she had chosen me.

When I stepped on board for the first time, I was accompanied by my tour guide, Mr. Bud Bronson, naval architect, Naval Academy graduate and longtime *Odyssey* skipper and volunteer.

Descending below decks was like stepping back in time. Dark wood, creaking floorboards and that old wooden boat smell—something like mustiness and turpentine, diesel oil and standing water in the bilge. The scent stays with the boat and leaves on anything that comes in contact with it.

As we made our way to the aft cabin, and master stateroom of the original owners, Barbara and Barklie Henry, Bronson told me to sit down at the little portside flip-top vanity where Barbara, a descendant of the once great Vanderbilt family, had sat probably countless times.

Besotted, I remained here for a moment, looking into the small mirror and imagining what her life must have been like, what their lives must have been like—nineteenth-century New York aristocracy well into the 1930s and 1940s, adventures to distant places and the coming of World War II.

I was hooked, and I knew that the *Odyssey* had many more tales to tell. Of trips abroad, espionage and famous friends. The kind of patriotism that lends itself to decorated service as a naval research vessel and top-secret projects we may never completely uncover. Of young naval recruits throughout the whole of the war and after, sailboat races with Bogart and, finally, as a training vessel for young adults. They have all had a place here. It's been an extraordinary journey to becoming Tacoma's Tall Ship to be sure.

I have stood at the helm of Tacoma's historic Tall Ship with the wind in my face during her time with Sea Scout Ship 190, where we have been merely her keepers, if only for a little while.

These are the stories of the *Odyssey* turned USS *Saluda* turned *Odyssey* once more—how she came to be and some of the lives that became intertwined and a part of her legacy. It's a legacy I have set out to preserve here in these pages by bringing the pieces of the story together, a story that belongs to us all.

Chapter 1

BEGINNING IN TACOMA

Under Paine's direction, Odyssey *was brought from "horrible condition" to certification by the Coast Guard with volunteer labor and services by many Tacoma businesses. Since then,* Odyssey *has become widely known on the waters of Puget Sound, serving as a floating goodwill ambassador for Tacoma as well as scouting itself.*

—Tacoma News Tribune[1]

It was 1978 when *Saluda*, formerly known as the *Odyssey*, came to Tacoma and remerged as Tacoma's Tall Ship. It was located at the Naval Air Station (NAS) at Whidbey Island, Washington, since 1974, and the navy had been trying to figure out what to do with the sleek 88-foot, 6¾-inch-long white sailboat that it had inherited during World War II, until fate stepped in.[2]

A little under two hours south of Whidbey Island on Commencement Bay lies the city of Tacoma, a vibrant maritime and sailing community on the shores of Puget Sound. It also was home to several chapters of the national and worldwide youth organization operating under Boy Scouts of America, called Sea Explorers at the time.

Unlike Boy Scouts, the widely unheard-of Sea Explorers (today called Sea Scouts) is a youth leadership program with a focus on maritime skills and water-based activities, with seamanship, camaraderie and teamwork at its core.

According to longtime supporters Terry and Renée Paine, it was Doug Cullen, a Sea Explorer skipper, who played a key role in the discovery and dissemination of the historic ship to the Mount Rainier Council of Boy Scouts in Tacoma (known as Pacific Harbors Council today).[3]

Leader of the *Corsair*, Sea Explorer Ship 448 (SES), Cullen and a few other ships were a part of the Mount Rainier Council region. On the lookout for a bigger boat, it came to his attention that there might be a sailboat on the navy's surplus list.[4]

Inquiries were made to the General Services Administration (GSA) in the fall of 1978. Although initially seeking something in the 40-foot range, Cullen was told all that was available was an almost 90-foot sailboat.[5] It was too large for his own needs and unique to most Sea Scout ships, which were notably smaller, but he saw a great opportunity and notified then Sea Explorer commodore Dave Phillips, who notified the council.[6]

They agreed that the ship could be an asset used by different scouting organizations in the region. A request was submitted by then Council Executive Director George Leonhard to obtain the boat. By December, she had made her way to Tacoma but was desperately in need of some TLC to bring her back to life and sailable again.[7]

That first year, she was moored at several waterways in the Port of Tacoma tide flats area. Finding dock space for an almost 90-foot boat with a 12-foot draft proved to be challenging more times than not, so it would be a while before she would find a more permanent home.[8]

One Sunday morning about a year later, Renée Paine recounts how she and her then fiancé, Terry Paine, were driving across the bridge to go to their own boat. Looking down at *Odyssey*, a bit worse for wear and tilting over a bit during low tide, was heartbreaking.[9]

"We saw this fabulous boat that looked like it had seen better days. As a consummate boater, Terry saw the potential and knew the quality that was there," said Renée. "She's bound with copper strapping. She's an oceangoing vessel. She's just fabulous, and he says to me, 'What a shame. Somebody should do something.' I told him, 'Why don't you?'"[10]

By Monday morning, having hardly slept, he was in the office of George Leonhard with a one-year plan to get the program up and running, and that's exactly what he would do.[11]

Paine was a teenager when he first began crewing, like Phillips, on another local Sea Explorer vessel named *Charles N. Curtis*. His skill as an attorney, community connections and can-do attitude was just what *Odyssey* needed since she was in such poor shape.[12]

The sails were ratty and patched. Long gone were the beautifully laid wooden decks, having been resurfaced with a nonskid layer covering all

Above: Dave Phillips, Sea Explorer commodore at the time, helped facilitate bringing *Saluda* to Tacoma. *Courtesy of Kent Gibson.*

Right: George Leonhard, Mount Rainier Council executive, was instrumental in bringing *Saluda* to Tacoma. *Courtesy of Kent Gibson.*

It was 1978 when USS *Saluda* left the navy at Whidbey Island, Washington, for Tacoma, to be used as a Sea Scouts training vessel. *Courtesy of Kent Gibson.*

Saluda was moored at several different waterways in the Port of Tacoma tide flats area. *Courtesy of Kent Gibson.*

About a year after *Saluda* came to Tacoma, it was Terry Paine who really got the *Odyssey* program up and running and made it what it is today. *Courtesy of Kent Gibson.*

but one of the naturally lit deck prisms and painted blue by the navy. The fireplace that once stood in the main salon was no more. There were water leaks and costly projects like repairs to the bow and stern, along with endless sanding and painting of the wooden hull and interior.[13]

What remained, though, were the beautiful lines Sparkman & Stephens is known for, the unquestionable quality of a Nevins-built boat and the many original features of this once luxurious pleasure craft.

From the salon's gimbaled table to the flip-top vanity in the aft cabin/master stateroom to features in the captain's cabin, galley and charthouse, *Odyssey*'s former glory still crept through even after all this time and disrepair.

It's what fueled the dream and vision of what she once was and could be again as the organization began to learn, too, about her important past. Soon, with the hearty embrace of the Tacoma community rallying behind them, and Paine leading the way, contributions to their cause would come. Recognizing the value of such a program to youth, and especially the maritime trade and tourism in their community, supporters in the Tacoma area and beyond gifted everything from funds to equipment, labor, parts and specialized services.[14]

One of the first things Paine did was to get her cleaned up and secured. He brought a few floating docks down so she was farther out and no longer running aground.[15]

Due to lack of upkeep by the navy, *Saluda* was looking very worn and needed a great deal of work by the time she came to Tacoma. *Courtesy of Kent Gibson.*

An early contributor named Samuel H. Brown got the ball rolling by investing funds for materials and the hiring of a shipwright.[16]

Paine went to a local vocational school where his friend was an instructor and brought the diesel class down to the boat to go through and make repairs. Tacoma Diesel & Equipment Company did a free overhaul of the diesel generator.[17]

Community organizations like the Tacoma Exchange Club stepped up by donating money to buy a diesel-fuel range for the galley and heater for the cabin.[18]

Concrete Technology Corporation in Tacoma donated use of a dry docking facility where the boat would spend many months on the hard undergoing necessary work.[19]

Another local business that donated its services was Parker Paint. According to Renée, its paint and professional painters spiffed up the hull above the waterline and painted the entire interior a glistening white for free, with not a drop out of place.[20]

Paine's efforts in the early days were instrumental in bringing the ship back to life. He not only established the *Odyssey* program but also created a sustainability model that lasted more than forty-five years. Weekend after weekend and month after month, die-hard volunteers young and old worked tirelessly. Although slow going, contributions from outside and thousands of man hours allowed the restoration to be completed in about a year.[21]

Returning to her original moniker, she officially became *Odyssey* once more at a public ceremony in May 1981. Many supporters from around the community were in attendance for the rechristening, including keynote speaker and Parker Paint sales manager Jack Fabulich. It was Sam Brown's wife, Natalie, who broke the celebratory champagne bottle on the bow.[22]

And so launched a high-adventure Tall Ship youth sailing program the likes of which had never been seen before on the Pacific Northwest coast. But it's some eighty-five years earlier when the story really begins, on the opposite coast.

Chapter 2

THE RISE OF THE *ODYSSEY*

She was christened in the fall of 1938 because I remember we were wearing overcoats at Nevins Shipyard in City Island, New York. My sister swung the champagne bottle against the hull and we slowly lowered down the ways. Aboard was my family, the professional crew, Olin Stephens, Drake Sparkman, Henry Nevins, with most of the others on the dock.

—William Barklie Henry, son of Barklie and Barbara Henry[23]

It was the halcyon days of sailing, according to a special feature in the Sunday edition of the *New York Times*, July 18, 1937—"Yachts Dot Island Harbors as Sport Rises in Popularity."[24] Sailing had become a favorite pastime, especially on New York's North Shore among those wealthy enough to support such pursuits.

Like their neighbors, the Henrys were very much a part of the Newport/New York/Palm Beach coastal yacht cruising and entertaining culture. Add in the occasional America's Cup and competitive regatta series races for prize money, trophies of gold and silver, bragging rights and spectating of the sport, and it is easy to see how this "golden age of sailing" had become a way of life.[25]

With Hempstead Harbor located just off Glen Cove and only five miles away from their home in Old Westbury, it was only a matter of time before Barklie and Barbara Henry would join the likes of the Morgans and the Woolworths with luxury yachts moored off New York Yacht Club's historic Station No. 10. Both Barbara's brother Cornelius Vanderbilt (C.V.)

Barbara Whitney Henry, pictured in March 1922 before her marriage to Barklie Henry just a few years later. *Author's collection.*

Whitney and father, Harry Payne Whitney, kept vessels here, along with many notable persons of the era.[26]

It was 1922 when she, Miss Barbara Vanderbilt Whitney, highly regarded young lady from the long-established Vanderbilt family, was debuted in society on New York's Upper East Side.[27] Two and a half years later, she and Barklie McKee Henry would marry in Roslyn, New York, setting in motion the extraordinary journey of Tacoma's Tall Ship, the *Odyssey*.[28]

She was the daughter of Mr. Harry Payne Whitney and Mrs. Gertrude Vanderbilt Whitney, and her great-great-grandfather was Cornelius Vanderbilt. Known as the Commodore, the nineteenth-century shipping and railroad tycoon was responsible for establishing one of the most prominent families in American history.[29]

Gertrude Vanderbilt Whitney was especially known in her own right as a sculptor, patron and collector of art, founding New York's Whitney Museum of American Art, which opened in 1931.[30] Harry Payne Whitney came from another notable family, inheriting much of his fortune from his father, William C. Whitney, a successful businessman and secretary of the navy under President Cleveland. The remainder of his wealth came from horses, both breeding and racing.[31] Along with homes in Manhattan and Newport and country estates on Long Island, Mr. Whitney owned a number of large sailboats and luxury motor yachts over the years.[32]

Although Barbara's father was a Yale man, her new husband-to-be graduated from Harvard.[33] The young Barklie was a product of successful Philadelphia banker William Barklie Henry and his wife, Alice Belknap,[34] and his father was a well-known yachtsman who served as rear commodore for the Corinthian Yacht Club in Philadelphia.[35] New York and Philadelphia were not that far apart, and the Henrys associated with many of the same wealthy families as the Whitneys.

The Gilded Age was at its close, but upper echelons of society like the Vanderbilt-Whitneys and Henrys still moved in elite social circles, traveled to Europe and had children raised by nursemaids and governesses. They wintered at grand Fifth Avenue apartments and entertained at exclusive parties, coming-out receptions and evenings at the opera. The summer

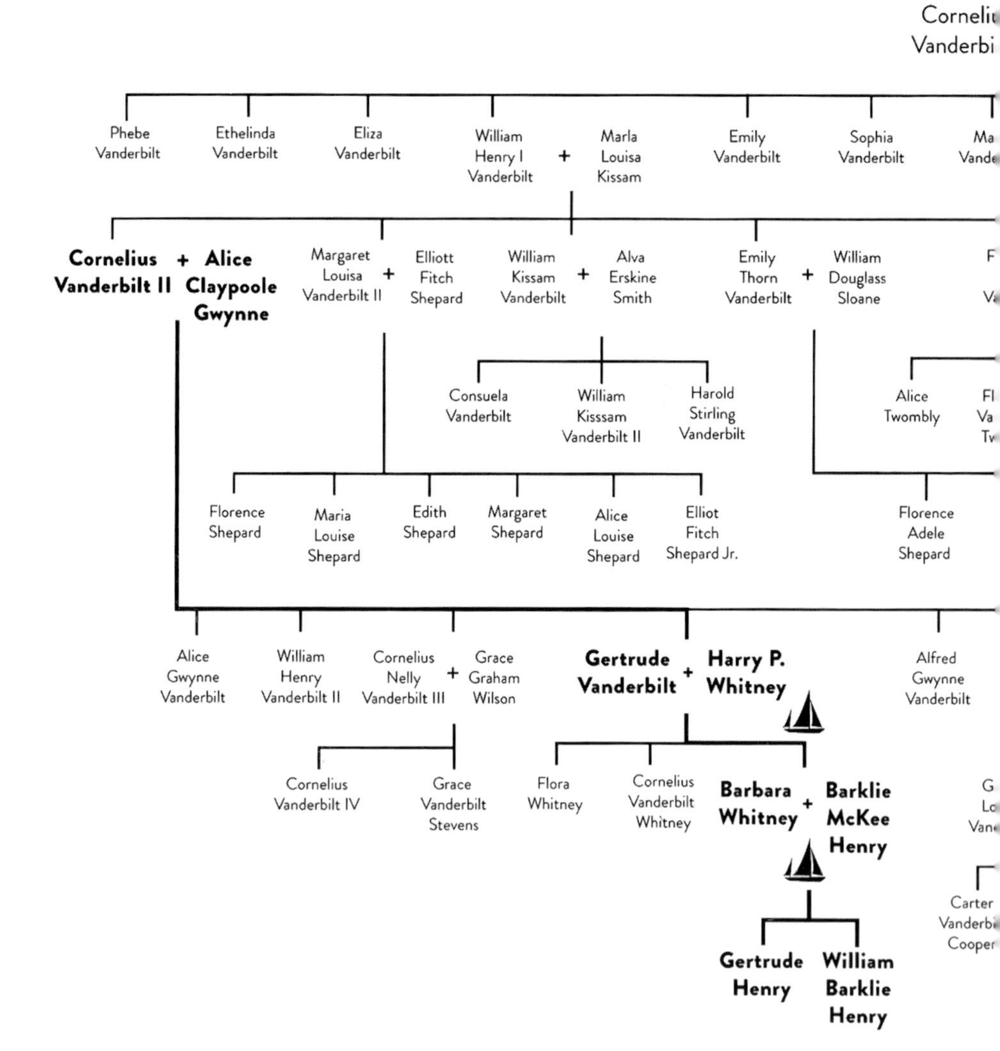

Vanderbilt family tree showing the line of descendants in relationship to the *Odyssey*. *Author's collection.*

ANDERBILT LY

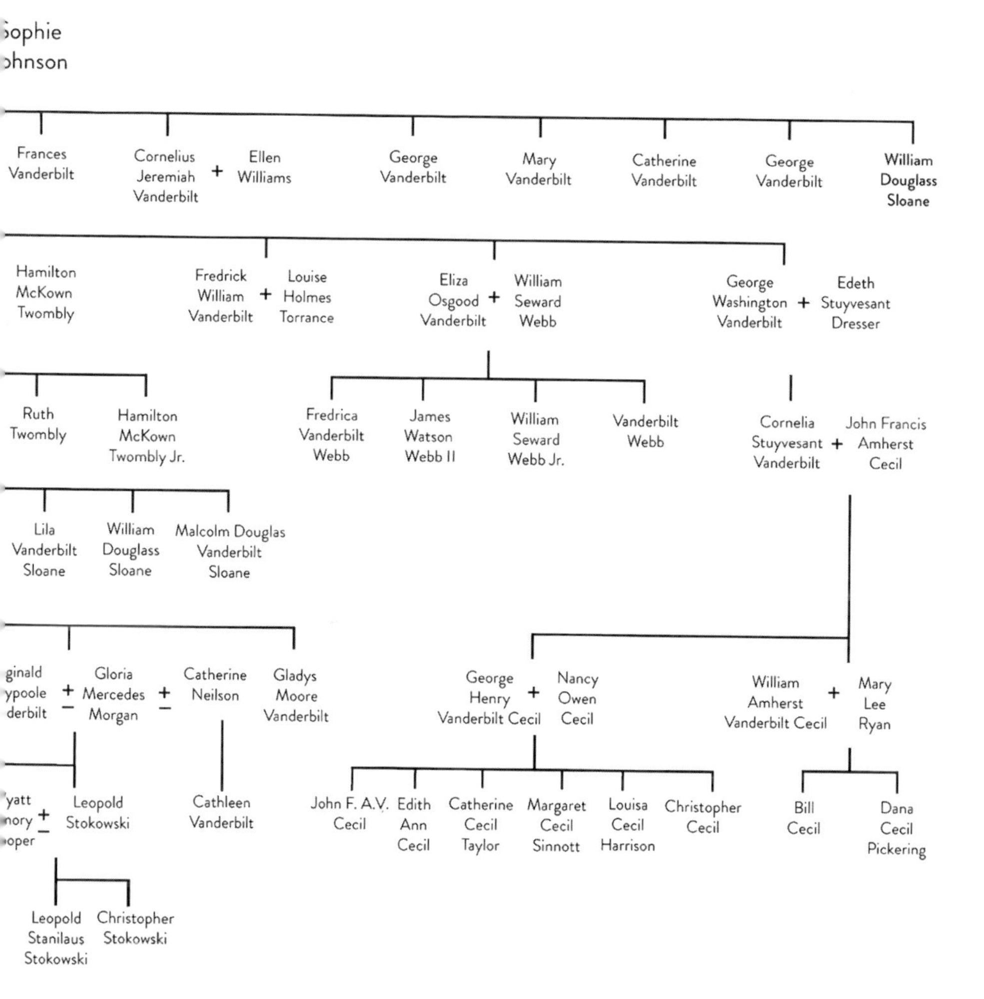

Left: Gertrude Vanderbilt Whitney in 1902, mother of Barbara Vanderbilt Whitney. Gertrude was a well-known artist and sculptor of her time. The Whitney Museum in New York is named after her. *Courtesy Whitney Museum of American Art, Gertrude Vanderbilt Whitney Papers, 1851–1975 (bulk 1888–1942), Archives of American Art, Smithsonian Institution.*

Opposite: Barbara W. Henry, pictured in 1935 with Louise O'Brian and Mr. Henry Shepard in Bermuda, where the Henrys often visited and had a home. *Author's collection.*

season was spent in Newport amid grandiose mansions and aboard lavish yachts up and down the eastern seaboard.[36]

It was a way of life each had, respectively, been born into, and so on June 25, 1924, the two were married at a Long Island church. A lavish reception followed at the expansive estate owned by the bride's parents in Wheatley Hills, called Whileaway.[37]

That first year of married life was spent abroad, beginning with travel across Europe. Finally, they set up house in Oxford, where the groom was enrolled in a special program. All accounts indicate that it was a time of wedded bliss. In just over nine months, Barbara would give birth to their first child in London, a daughter called Gertrude, born on March 27, 1925.[38]

Around 1926, the little family returned stateside, dividing time between several properties. Two were located not far from Barklie's alma mater in Boston, as well as a short distance away in Manchester, New Hampshire.[39]

It was also during this time that Barklie managed to pen a novel titled *Deceit*. The book described the inner social workings of the Boston, Newport, New York and Palm Beach set.[40] This was the very set to which he and Barbara belonged and most likely how they became acquainted.

Barbara fit into her role as high society wife and mother like the ladies she'd known before her. Keeping up appearances of a prosperous household, entertaining and leisure pursuits were full-time duties.[41]

Barklie, on the other hand, worked at the *Boston American* and *Atlantic Monthly*. He later became the managing editor of the *Youth's Companion*, a popular Boston publication among young people of the time.[42] Positions in the literary sphere, authoring his own book and having married into one of the wealthiest families in America at the time helped him form some pretty famous connections, like one with Ernest Hemingway. The two maintained a correspondence between 1926 through 1929 and well after.[43]

Barklie McKee Henry went to Harvard, later moving in literary spheres with the likes of Ernest Hemingway before entering a banking career like his father. *Courtesy of Linda Barnwell.*

According to one letter dated July 14, 1927, it was Barklie, known as "Buz" by Hemingway and friends, who was helpful in getting Hemingway published in the *Atlantic Monthly*: "How are Barbara and your family? I sent you a wire when your first child was born but haven't read of any others in the Spanish Austrian or French papers so I don't know to what extent you've become a father. It was swell of you to speak to Mr. Sedgewick and I think it was cockeyed sporting of them to publish the story. I was terribly pleased you liked it. Will have it in a book along with 12 others this fall. I'll write Scribner's to send you one."[44] Owner-editor at the *Atlantic Monthly* Ellery Sedgewick became one of the first American editors to publish Hemingway's work.[45]

Buz also facilitated a friendship between Hemingway and another American writer, Owen Wister, by sharing Hemingway's early collection of short stories *In Our Time* with him. This series preceded his great success with *The Sun Also Rises*.[46] Barklie eventually would leave the writing world behind, but he kept up a friendship with Hemingway over the years, as confirmed by the family in later communications.

The couple's second child was born on November 18, 1927, William Barklie Henry, named after Barklie's father.[47] Like his father, Barklie would take a position in banking, eventually moving the family to New York, where, in 1928, he worked in the bonds sector at the Guaranty Trust Company.[48]

Mr. Henry rose to president of the Association for Improving the Condition of the Poor, a trustee at Cooper Union and a council member at New York University. Along with numerous other director and committee roles, he became president at New York Hospital and director of Texas Company and the U.S. Trust Company.[49]

By the time he was thirty-seven, he was a director at another rising industry of the day, American Telephone & Telegraph Company (AT&T).[50] These and other lucrative ventures led the family of four to a more permanent

Long Island residence suitable for raising children. The stunning twelve-acre Noel & Miller–designed estate, built in 1930, was located in Old Westbury.[51]

Surrounded by elaborate estates, country clubs and private schools, it was a segue to the prominent Long Island sailing culture. Although yachting had always played a role in each of their lives, it would soon touch the lives of their own little family as they remerged within the popular yachting community of nearby Glen Cove.[52]

CHAPTER 3

THE BIGGEST THAT YEAR

NEW YAWL IS LAUNCHED
MRS. HENRY'S YACHT CHRISTENED ODYSSEY *GOES INTO WATER*
Taking the water easily yesterday morning on the high tide, Mrs. Barklie Henry's new yawl Odyssey, *designed by Sparkman & Stephens, and built by Henry B. Nevins, was launched at City Island.*
The new offshore yacht, too big for the Bermuda race, has a conventional rig, but with towering mainmast over 104 feet long. It is 95 feet above deck or 12 feet more than Baruna, the last Bermuda winner.
Work of rigging the new craft will start tomorrow. When that is finished, she will start south. Mrs. Henry, a daughter of the late Harry Payne Whitney, planning to cruise in the West Indies this summer.
—New York Times, *October 23, 1938*[53]

In the spring of 1938, it was announced that "Mrs. Barklie Henry's Sparkman & Stephens designed yawl, christened the *Odyssey*, would be ready for Long Island Sound early that summer."[54] Although Mr. Henry would be very involved throughout the design process, it was rumored that the new yacht was to have been a birthday gift for him, purchased by Mrs. Henry.[55]

The name harkened to his writing and literary days. A lover of the classics, Latin, ancient Greece and particularly Homer, he drew from these in naming the vessel. "The *Odyssey* tale was close to his heart and it was no accident that the boat he spent so many hours detailing was named after his work," said William Barklie Henry, son of Barbara and Barklie Henry.[56]

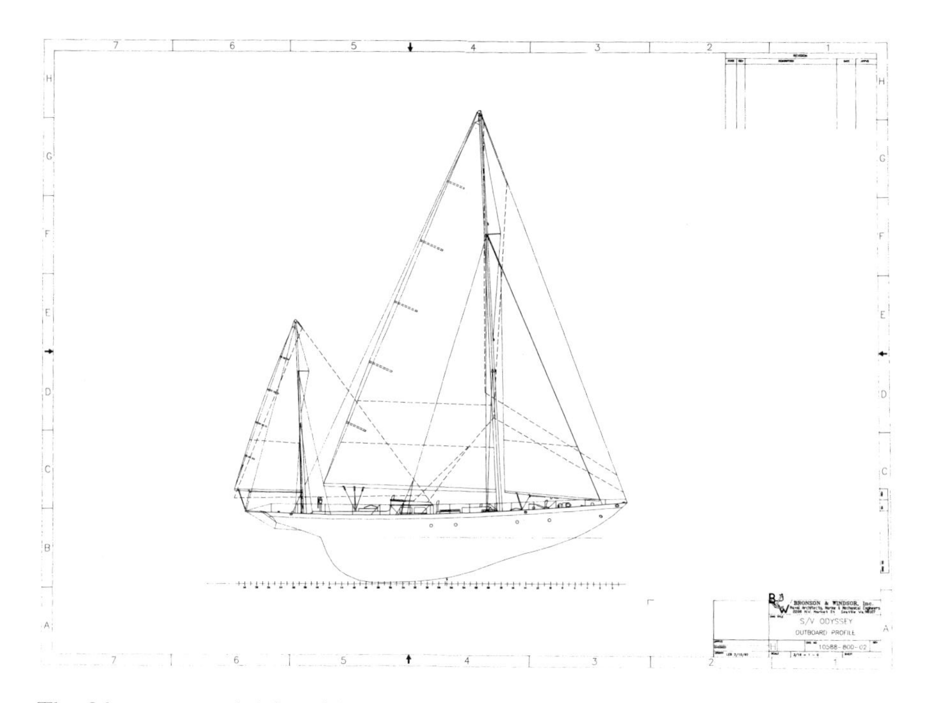

The *Odyssey* measured eighty-eight feet, seven inches in length, with an eighteen-foot beam and a ten-foot, eight-inch draft; she had fifty thousand pounds of lead ballast in her keel and was one of the largest vessels to be constructed that year on City Island, New York, at the Henry B. Nevins Shipyard. *Courtesy of Bud Bronson.*

To design their yacht, the Henrys chose the firm of Sparkman & Stephens (S&S), well-respected yacht designers for the better part of the twentieth century. The quintessential partnership, created by brothers Olin and Roderick "Rod" Stephens Jr. and Drake Sparkman, was second to none. A number of wooden S&S yachts took the world by storm between 1929 through the late 1930s.[57]

Both Stephens brothers were expert yachtsmen with noteworthy wins aboard designs like the 1930 52-footer *Dorade*, which won the Transatlantic in 1931.[58] *Odyssey* is said to have taken much of its inspiration from the similarly appointed, but smaller, *Dorade*.

Their designs would produce numerous America's Cup winners, one of which came from yet another Vanderbilt, renowned National Sailing Hall of Famer Harold S. Vanderbilt (first cousin to Gertrude Vanderbilt Whitney). The design, the 130-foot *Ranger*, was a collaboration with another distinct yacht designer, W. Starling Burgess.[59]

According to William Barklie, who spent many days aboard *Ranger* in Newport, "Interestingly, Harold Vanderbilt married my father's half-sister, Gertie. I saw them a lot as a child."[60]

Completing the *Odyssey* trifecta was Henry Nevins, wooden yacht builder for what can only be described as the pinnacle of both wooden boat design and wooden boat construction. These were the uppermost crest of boatmaking, experts in their own right and regarded as so still today.

Nevins Ship Yard in City Island, New York, became a renowned wooden boat builder in the early 1900s and remained so through the Depression years until well after Nevins's death in 1950.[61] The man himself was said to have been very humble even after making his millions. He frequently carried his lunch to work and ate with the very workers he employed, knowing each by name.[62]

While Olin Stephens was founding the firm with Sparkman, it was the Nevins boatyard where Rod Stephens got his start before joining the other two.[63] Together, Nevins and S&S would have many winning combinations throughout their lengthy careers, so many that even Olin Stephens couldn't tell them apart at times.

In the 1960s, William Barklie designed a new sailboat racing rig. He used the original tank-test hull model of *Odyssey* since the five-foot-long model was similar to the form of his rig. It was originally used by Olin Stephens when testing *Odyssey*'s hull in 1937. His father had taken him to the only testing tank available at the time, located in Hoboken, New Jersey, to witness the event.[64]

Anxious to share his new design with the esteemed Olin Stephens, he made an appointment at his New York City office. "I remember the model almost filling the elevator up to his office," recalled William Barklie.[65]

After Mr. Stephens fondly remembered his mother and father, he looked the model over while discussing the novel rig, noticing that it was a tank-test model due to a strip of sandpaper on the bow. "He asked me what boat it was," described William Barklie. "I answered, 'It's one of yours. Can you identify it?' He said, '*Vim*.' And I replied, 'No, it's *Odyssey*,' and he answered, 'Just about the same.'"[66]

Vim was another sailboat made for Harold S. Vanderbilt by the trio, the first 12-meter America's Cup defender. William Barklie said of the meeting, "I think it's interesting that America's best sailboat designer for many years mistook the hull of a 90-foot cruising yawl for a 65-foot racing sloop."[67]

With World War II fast approaching, both outfits went on to play an important role for the U.S. Navy—one designing amphibious troop landing

Odyssey's maiden voyage took place off Nevins Shipyard, City Island, New York, in 1938. *Courtesy of SSSOdyssey.org.*

craft and the other building minesweepers and aircraft rescue boats.[68] S&S even won a U.S. Medal of Freedom for its collaboration with GMC on the DUKW series six-wheel water-going vehicle known as "duck."[69]

Nevins had the distinction of building the first auxiliary motor minesweeper, the YMS series. The design plans were completed in many a boat yard, and it was an integral asset in clearing mines laid by enemy submarines during the war.[70]

Spring passed and soon summer, until finally the culmination of the trio's work for the Henrys, #433, *Odyssey*, was completed later that fall.

Odyssey was launched on October 23, 1938, to much fanfare off Nevins boatyard in the Bronx. Gertrude, with her brother nearby, was the one to christen the ship by successfully breaking the traditional bottle of champagne over her bow. A number of guests turned out to see the new yacht as it took to the tide. All that remained was rigging of the craft and the planning of great and distant voyages.[71]

Captain Oren McIntyre, Gertrude Henry and William Barklie Henry pictured together at the launch of *Odyssey* in 1938. *Courtesy of SSSOdyssey.org.*

Measuring 88 feet, 7 inches in length with an 18-foot beam, a 10-foot 8-inch draft and fifty thousand pounds of lead ballast in her keel, she was one of the largest to be constructed that year.[72] The double-planked white cedar and Honduran mahogany hull, with its 104-foot mainmast towering high above, boasted a sail area of 3,453 square feet. The teak decks, rails, trim and butternut throughout the interior distinguished *Odyssey* as a hallmark of quality and craftsmanship.[73]

Mr. Henry put a great deal of time into finding a wood carver to create the immense reliefs depicting Homer's *Odyssey* that were installed in the main salon.[74]

Completely state of the art at the time, she was equipped with an eighty-five- to one-hundred-horsepower MRA-6 Superior Diesel marine engine, cutting-edge radio and direction finder, modern plumbing and a GE refrigerator.[75]

The roomy deckhouse was a perfect combination of portside dinette and chart table to starboard. This is where the navigational charts and two-way radio and direction finder equipment were housed.[76]

Left: William and Gertrude on the bow of *Odyssey* as Gertrude gets ready to break the champagne bottle, customary for good luck. *Courtesy of SSSOdyssey.org.*

Below: Gertrude, daughter of Barbara and Barklie Henry, christens the ship with bottle of champagne on her bow, 1938. *Courtesy of SSSOdyssey.org.*

Left: Gertrude christening *Odyssey*, 1938. *Courtesy of SSSOdyssey.org.*

Below: *Odyssey* dressed with flags off City Island, New York, during maiden voyage, 1938. *Courtesy of SSSOdyssey.org.*

Opposite, top: Interior photos featuring *Odyssey*'s charthouse/navigation room. This is where navigation of the ship takes place; charts (maps) and radio equipment are kept here still today. *Courtesy of SSSOdyssey.org.*

Opposite, bottom: Shots of *Odyssey*'s interior when she was brand new, featuring the fireplace, which would later be removed by the navy. Today, the spot is used as a small closet. *Courtesy of SSSOdyssey.org.*

Photos of *Odyssey*'s interior featuring the gimbaled table (which always remains level even when the ship is moving). The same table is still on board today. *Courtesy of SSSOdyssey.org.*

Odyssey's main salon was the height of luxury and modernity in 1938. *Courtesy of SSSOdyssey.org.*

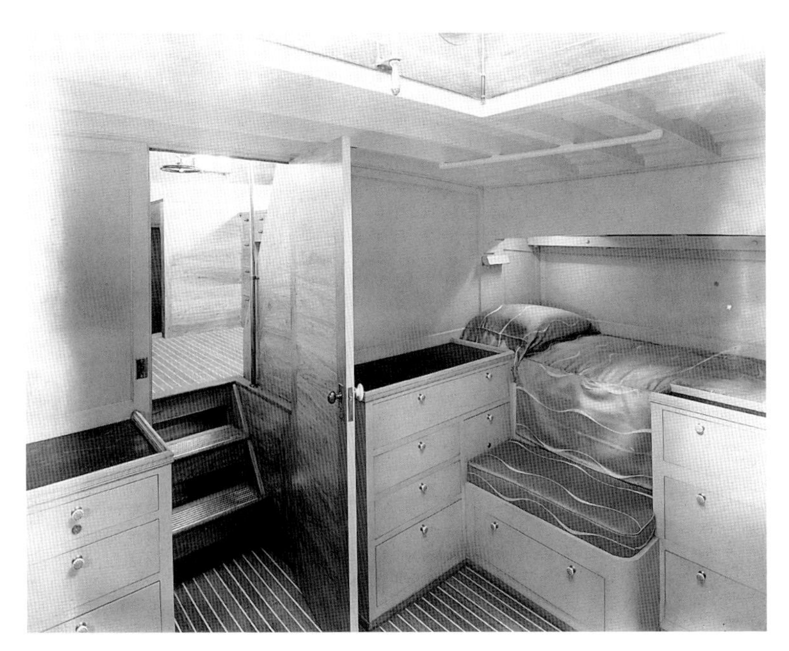

Interior photo of the aft cabin, aka the main stateroom. This is where Barbara and Barklie Henry slept. On the port side (left) of ship was Barbara's bed, next to flip-top vanity. Everything is set up the same today. *Courtesy of SSSOdyssey.org.*

Odyssey's state-of-the-art galley was the height of modernity at the time and even came equipped with a refrigerator and freezer. *Courtesy of SSSOdyssey.org.*

Just forward and a few steps down was the main cabin common area, often called a saloon or salon. It served much like a living room on board the vessel, complete with fireplace and piano.[77]

Along with a spacious aft stateroom, with its own private head (or bathroom) boasting a bathtub and a shower for Mr. and Mrs. Henry, another stateroom was situated amidship for the Henry children, directly across the passageway from a second head.[78]

The captain's cabin was near the well-fitted galley, replete with a peerless refrigeration vault. Forward of that was the forecastle and where the rest of the working crew slept, separated, of course, from the Henry family by a door aft of the galley—they would climb up the ladder and through the forecastle hatch to come around to carry out duties.[79]

Plenty more spaces could be converted into berths if needed, and everyone would have their place on board as they readied for the journeys ahead.[80]

All this for a pretty penny too. The total cost to build and outfit her came in around $87,669.66, with a price tag of $96,379.50.[81]

Chapter 4

THE *ODYSSEY* SETS SAIL

Julys the family sailed along the coast of Maine on my father's white yawl. It was often stormy and foggy and grey. When I heard the moaning of the stays I was afraid of letting go. But then all of a sudden the sky would clear and little islands would appear, bright and shining. Down went the anchor in a protected cove. In my bunk I would read Little Women and listen to the grownups on deck talking and laughing.[82]

—Gertrude Vanderbilt Whitney Conner, daughter of Barklie and Barbara Henry

The family were finally in possession of their new cruising yawl, the slick white-hulled beauty named *Odyssey*. Moored off Glen Cove, this is where she remained in between shorter sails around the New York–New England coastal region. Plans were underway for large-scale cruises that would take place that winter. All she needed now was a crew.[83]

Although sailors the lot of them, typically families of the Henrys' means had paid crewmen to maintain and sail their boats for them. They were lucky to have experienced longtime Maine sailor Captain Oren McIntyre in their employ. A veteran skipper for more than twenty years with a penchant for anything with sails, he even had writings about the subject published in the *Saturday Evening Post*.[84]

McIntyre had been at the helm of smaller vessels owned by the family and, now, aboard the *Odyssey*. The four remaining crewmen consisted of Carl Roed; Teddy, the cook; and a German sailor named Sandy Schmidt.[85]

Interestingly, Schmidt, who was known to have excellent penmanship, was required daily to copy notes made by Captain McIntyre and Mr. Henry into the official ship's log. This resulted in a beautifully written and legible log that reads like a storybook.[86]

Several more sailors were associated with joining or replacing crewman over the next few years, including Billy Olsen, Harry Fontaine and Jerry Towhill, but McIntyre was always in command as the captain with at least four to assist him.[87]

"Our first sail was out of her mooring at Glen Cove. Our first cruise was over Christmas vacation that year and we met the boat in Miami. Our itinerary for the two-week cruise was, Florida Keys, Cuba, and return to Miami. I have a fairly clear memory of visiting Tortugas where we were the only visitors to tour the fort that day. We also stopped at Key West where we had lunch with Ernest Hemingway," said William Barklie.[88]

According to details by William Barklie, it was the winter of 1938, and the trip would last over Christmas and New Years. The crew sailed *Odyssey* ahead of the family, who arrived later by airplane. Regrettably, it was one of the first and very few trips the family would take on her because the war came along soon after.[89]

These early experiences are what helped William develop a strong love of boating, unlike for his sister. The memories on board the *Odyssey* clearly remained with him throughout the rest of his life. From he and Gertrude shining the brass on board to other little jobs they were given, he remembered the time with his family fondly.[90]

The journey, which lasted two weeks, embarked from Miami, traveling onward through the Florida Keys, to Cuba and back to Miami again.[91] There were twelve people on board, including the crew of five, a doctor friend of the family, the Henrys and their two children and then another couple, "I think my sister and I slept in the main cabin, and the couple slept in the forward cabin. My parents slept in the aft cabin. I can't remember where the doctor slept, and of course, the crew had their own spaces," said William Barklie.[92]

One of the stops the *Odyssey* made was Key West, where the family joined Ernest Hemingway at his home for lunch. Hemingway had purchased a house at 907 Whitehead Street in the early 1930s with his then wife, Pauline.

Opposite: Close-up featuring Captain McIntyre. *Courtesy of SSSOdyssey.org.*

Left: Captain Oren McIntyre pictured at the helm of *Odyssey*. *Courtesy of SSSOdyssey.org.*

Below: *Odyssey*'s first crew, with experienced Maine sailor Captain Oren McIntyre at the helm. Captain Mac was employed by the Henry family on other vessels owned by the family even before *Odyssey*. *Courtesy of SSSOdyssey.org.*

BARKLIE HENRY
31 NASSAU STREET
BArclay 7-0976

TO: Captain Oren F. McIntyre SUBJECT:

October 9, 1941.

<u>ODYSSEY EXPENSES for first nine months of 1941</u>:

	Actual	Budget	(Over) or Under
Food	1,971.69	3,000.	1,028.
Shore Board (Subsistence)	756.00	750.	(6.)
Wages	6,089.06	6,300.	211.
Social Security	52.78	57.	4.
Uniforms	270.21	225.	(45.)
Insurance	257.50	2,625.	2,368.
Fuel	476.48	600.	124.
Laundry	509.22	600.	91.
Galley & Cabin Supplies	251.26	300.	49.
Navigating Accessories	106.97	150.	43.
Shipyard Expenses	3,491.19	3,000.	(491.)
Deck Dept.	179.64	750.	570.
Engine Dept.	-	750.	750.
Sailmakers' Bills	3,163.00	2,400.	(763.)
Sporting Goods	12.85	150.	137.
Port Charges & Pilot Fees	301.04	150.	(151.)
Guides & Boat Hire	140.50	75.	(66.)
Radio Telephone	149.76	390.	240.
Miscellaneous	310.47	150.	(160.)
Total Odyssey expenses	18,489.62	22,422.	* 3,932.

*The above shows that you are $3,932. under for the period. Please keep in mind that there is a large insurance bill coming due this month which changes the picture somewhat, making the total under only about $1,000.

Above: An *Odyssey* expense bill from October 1941, showing costs paid out for crew wages, sails and so on. *Courtesy of SSSOdyssey.org.*

Opposite: Page shown from *Odyssey*'s beautifully handwritten log book. Captain Mac and Mr. Henry had German sailor Sandy Schmidt rewrite their notes into the log because of his very neat handwriting. Dated August 25, 1941. *Author's collection.*

YACHT "ODYSSEY" FROM Marblehead Mass [illegible] DATE Aug 25 1941 5

PLACE TO

TIME	SEA	WIND AND FORCE	WEATHER	COURSE	REMARKS
A.M. 1					This day comes in Cloudy
2					light Southerly breeze
3					
4					
5					Crew employed in general work
6					
7					
8					
9					Showers for this afternoon
10					
11			Fog		
12					
P.M. 1					
2					
3					Mr an Mrs Henry arr. tonight at
4					9.30 P.M
5					
6					Weather fresh Southerly breeze with rain
7					
8					
9					So ends this day
10					arrived with Posie & Buz.
11					Boat's wide awake [illegible]
12					St. J. Stonington & Two friends from Marblehead aboard.

TOTAL TIME UNDER WAY: ________ HOURS TOTAL DISTANCE MADE GOOD: ________ MILES

GENERAL: [LOSSES? DAMAGES? ILLNESS? PERSONNEL CHANGES? GUESTS? UNUSUAL INCIDENTS? FISHING LUCK?]

Where special report to owner or insurance company may be necessary, use next page to describe circumstances in full. SIGNATURE________

The home, surrounded by lush floral gardens, had undergone massive renovations and today is a public museum.[93]

When the Henrys arrived, they were seated at an outdoor table under a large tree. They could see him at the far side of the garden sparring with a very large Black man.[94] This may have been Shine Forbes, a local boxer Hemingway had befriended and paid to be his sparring partner. For several years before the addition of a mammoth pool, Hemingway had a boxing ring in his yard in the very spot.[95]

As introductions were made, William Barklie noted that Hemingway, still sweaty from the exertion and wearing boxing gloves, proceeded to go and take a shower. This made him late for lunch, as the family were served the first course without him.[96]

Besides the literary connection between Hemingway and Mr. Henry ("Buz"), they also would have fishing, wild game hunting and boating in common. Hemingway, who owned a much-loved 38-foot wooden fishing boat named *Pilar*, spent many days aboard his yacht customized in the Wheeler Shipyard in Coney Island, New York.[97]

Odyssey pictured out on the waters off the eastern seaboard, most likely motoring since no sails are being used, but there's a wake from ship's movement. *Courtesy of SSSOdyssey.org.*

Odyssey's journey continued, leaving Key West and crossing over to Havana. "I remember being excited about the comfort and power of *Odyssey* under sail in a fairly big sea," said William Barklie.[98]

The party visited Havana for five days, taking in the sights during a series of short excursions. One of these included going to see the home of William Barklie's grandmother Gertrude Vanderbilt Whitney, which wasn't far from

Odyssey at sail. *Courtesy of SSSOdyssey.org.*

Hemingway's house there. William Barklie remarked, "At eleven years old, I remember thinking that Hemingway must be really successful for having two really nice houses in such beautiful locales."[99]

A funny story recounted from the Havana trip happened on the morning of New Year's Day. William Barklie was awakened by his father on deck screaming at Captain McIntyre, who had returned rather late after being

Another shot of *Odyssey* at full sail, with all her sails. *Courtesy of SSSOdyssey.org.*

granted shore leave the night before. "My father rarely lost his temper, but when he did, it was spectacular," said William Barklie.[100]

Mr. Henry, who had expected McIntyre to return to the yacht before midnight, had encountered a rather close call with another large ship at two o'clock in the morning. As the vessel maneuvered into dock, it had come very close to the *Odyssey*, causing several horn blasts to ring out and *Odyssey* to spin around at anchor from the force of its prop wash.[101]

This awoke Mr. Henry, who had no doubt done a bit of celebrating of his own the night before, causing him to rouse the crew and notice its captain unaccounted for. "I had slept through the confusion, but not Captain McIntyre's dawn dressing-down," said William Barklie.[102]

Chapter 5

THE GALAPAGOS ISLANDS

My next cruise on Odyssey *was undoubtedly her major cruise with our family. It was around February or March of 1940. My sister and I were taken out of school for six weeks, and flown down to Panama to meet the boat. Our itinerary was through the Panama Canal, Cocas Island, Galapagos Islands, and the North Coast of Panama.*
—William Barklie Henry[103]

The Galapagos trip, and the Henry family's longest voyage aboard, began on February 11, 1940. The *Odyssey*, with American flags painted on her hull, must have been a sight to behold as the crew made their way from the Florida coast and through the Panama Canal to meet the party in Panama.[104]

The Henrys, joined there by three other guests, arrived by plane to embark on the journey ahead. Accompanying them were Montana dude ranch owners Fran and Harry Alderson[105] and a young doctor named Harold Genvert from New York Hospital, where Mr. Henry was president at the time.[106]

The first in a series of adventures along the way was a rite of passage that occurs when crossing over the equator, known by centuries of seafarers and celebrated in different ways. The *Odyssey* and guests got to join in the time-honored tradition of Neptunus Rex, the God of the Sea, as evidenced in photos and certificates passed down from the occasion.

Odyssey is seen here with the American flag, which the family had painted on both sides. In the pre–World War II timeframe, while sailing on a long journey, they wanted it to be known that this was an American vessel. *Courtesy of SSSOdyssey.org.*

Indoctrinated into the special club, all aboard received a cold shower of sea water and certificates stating the following: "The realm of his exalted majesty Neptunus Rex, and shall now and hereafter bear the title 'Exalted Shellback….' This was a name given to those who crossed the equator for the first time, while prior to this they are referred to as 'polliwogs.'"[107]

"The Galapagos Islands were extraordinary for an 11-year-old boy who enjoyed nature and loved to explore," said William Barklie of the trip. He and Gertrude, who was thirteen years old at the time, were taken out of school for the journey. His father had done a great deal of research among his yachting friends and connections at the New York Natural History Museum prior to leaving.[108]

"It was a great adventure, and the Galapagos was very virgin. Not that many people went there then. We really got to explore it," said William Barklie. "We did a lot of fishing, went ashore a lot, and it was a really marvelous trip."[109]

Though mostly for pleasure, the journey was also an opportunity for Mr. Henry to engage in a bit of medical philanthropy for the benefit of Johns

Barklie Henry being sworn in by Cap McIntyre and "accepted into the realm of his exalted majesty Neptunus Rex," to now and hereafter bear the title "Exalted Shellback." This was a name given to those who crossed the Equator for the first time, while prior to this they were referred to as "polliwogs." *Courtesy of SSSOdyssey.org.*

Baptism by cold water was part of the ceremony, seen here provided by Captain Mac to young William and his father, Barklie Henry. *Courtesy of SSSOdyssey.org.*

Barbara Henry is sprayed with cold sea water by Cap Mac (who seems to be amused) as part of the equator-crossing tradition. A crew member at the helm looks on. *Courtesy of SSSOdyssey.org.*

Barklie Henry with his children, Gertrude and William, in the Galapagos Islands, seen with a sea mammal. It was *Odyssey*'s major cruise with the Henry family. *Courtesy of SSSOdyssey.org.*

Above: William Barklie Henry pictured as a young boy during *Odyssey*'s great voyage to the Galapagos Islands. *Courtesy of SSSOdyssey.org*

Hopkins Hospital in Baltimore.[110] The "mission," as William Barklie called it, included the collection of more than one hundred specimens of fish eyes to be used in the name of science.[111]

Cases of jars and gallons of formaldehyde were carefully stored beneath the floors of the main salon for safe keeping until the return trip.[112]

William Barklie was happily charged with removing the eyes and placing them in labeled jars. The largest specimen came from a giant manta ray, often called Giant Devil Ray or Devil Fish. It had a wingspan of twenty-two feet, and although William was excited to help with the project, he was sad to see it harpooned by Captain McIntyre.[113]

He noted at the time that his parents were game fisherman, a practice he seems to have held a differing opinion of. One such anecdote was how his father and Captain McIntyre routinely set a shark line off the ship's stern, usually at night during the length of the cruise. This seems to have been a common practice among yachtsman during the era, according to William Henry.[114]

Using an enormous hook with a steel bar and chain leader, they would bait it with recently caught fish and hang it off the back of the boat at bedtime. This would naturally create quite a commotion, and the young William Barklie would awaken and sneak above decks to see what was happening.[115]

Above: Reeling in a large manta ray. Mr. Henry was collecting eye specimens of various sea creatures for research at Johns Hopkins. William talked about helping with the collection and storage. The manta ray was the largest of their specimens. *Courtesy of SSSOdyssey.org.*

Right: Barbara Henry at the barrel-shaped mailbox at Post Office Bay in the Galapagos, where early mariners started a tradition of leaving mail that others would deliver. *Courtesy of SSSOdyssey.org.*

Captain McIntyre, William and crew as a shark is caught in the shark line off the stern. Apparently, it was commonplace to set a shark line. *Courtesy of SSSOdyssey.org.*

Toward the end of the cruise near the Panama Coast, he saw a ten-foot tiger shark attempting to break free of the line. Two more shark heads pushed up the chain above the hook, and lots of blood was drawing more sharks to circle around. It was all quite shocking to an eleven-year-old boy and left a very clear impression even years later.[116]

Some of the less shocking activities enjoyed by the travelers were fishing, taking photographs and birding.[117] Harry Alderson, the rancher in their party, provided a great deal of entertainment, as it appeared he could lasso about anything. He always set it free afterward. "I remember he lassoed penguins, and even a seal," said William Barklie. "It was the craziest thing, seeing this guy swinging this rope around over there lassoing iguanas, and various things."[118]

They toured several of the islands that make up the archipelago, and one such excursion took the group to Floreana Island, also called Santa Maria Island, situated on its south end. Here they went in search of "the cave of the baroness," the setting for a tale based on an island love triangle—a murder mystery resulting in the possible death of two men.[119]

It was relayed to the *Odyssey* party that President Roosevelt's ship had stopped at the very same spot in search of the cave during one of his recent cruises. No one knows whether they found it, but the Henry group did. After locating the cave, they were shown a grave that was said to be that of the Baroness Eloise Wagner de Bosquet herself.[120]

Another nearby stop was Post Office Bay. Early whaling ships at sea for lengthy periods of time would stop here. Without any means of communicating with loved ones, they began leaving letters that would be delivered on by other sailors passing through.[121]

The custom, one that continues still today, is to search out letters addressed to anyone who lives nearby and deliver them on the return trip with an explanation of the journey. Of course, many people are known to simply place a postage stamp and mail it, but so the tradition goes.[122] The *Odyssey* crew brought back a stack of letters from the barrel-shaped mailbox to deliver accordingly.[123]

Although the journey would come to an end just before the Panama Canal, all but the crew disembarked for the return trip home by airplane. The crew prepared to sail onward to Miami's Merrill Stevens Boat Yard for a haul out before making the trip back to *Odyssey*'s homeport in New York.[124]

All seemed to go well except the short four hundred miles between Cape San Antonio, Cuba, and Miami, where it was slow going against strong headwinds that turned the usually short trip into five days, while ravaging both a jib and a staysail. "It was as tough of going as anything I have ever seen," Captain McIntyre had commented.[125]

Finally safe at port, McIntyre divided his time between the boatyard and his Miami home, directing the crew in general ship maintenance for the rest of that spring. *Odyssey* would set sail for New York some weeks later.[126]

After the family returned home to Glen Cove, the remainder of that summer and henceforward saw the Henrys sailing and entertaining mostly on local waters, although they would travel the length of the

A rare photo of Barbara Henry at the helm of *Odyssey*. *Courtesy of SSSOdyssey.org.*

eastern seaboard and particularly the coast of Maine, a familiar territory for McIntyre.[127]

Pages from the final ship's log between August 1941 and January 1942 recount cruising sojourns as far north as Maine and as far south as Florida.[128] Reading like a diary of sorts, the log chronicles everything from the names of specific visitors; when they came aboard; when they went ashore for lunch; when they went fishing, played golf or had dinner; along with the weather, wind and conditions of the sea.

There were stops all along the Massachusetts coast at places like Marblehead and Cape Cod. There were trips to Pemaquid Point and anchorages at Boothbay Harbor and Christmas Cove in Maine.[129]

The *Odyssey* traveled through the Cape Cod Canal and onward to Block Island, Rhode Island. There were frequent runs between Glen Cove, City Island and the whole of Long Island Sound in New York.[130]

But as the Henrys' time with the *Odyssey* was coming to a close, so, too, were the winds of change shifting for the rest of the world, as America would soon find itself on the brink of World War II. "Most of the time my father was not on board during this period, because he, like some of his yachting friends, had joined the Naval Reserves," said William Barklie.[131]

Officially, the formation of the naval reserves happened in 1915 during World War I. By the time World War II rolled around, reservists outnumbered active sailors in the navy.[132] "By the time we entered the war, he was a lieutenant commander with his own minesweeper, and went on to sub-chasers, and finally destroyers. By the end of the war, he was head of the sub-chaser training school in Miami."[133]

As the *Odyssey* sailed south that fall, the log chronicled preparations by the crew, such as this one for October 27, 1941: "This day comes in hazy, SW wind. Crew employed at general work, and getting ready to go to sea. Got underway at 1:30pm for Miami, FL. We had rain off and on. So ends this day."[134]

During this trip and many of the others, the captain (or crewman Sandy Schmidt from earlier) made notations of the landmarks they passed along the way, such as Cape Hatteras Lighthouse, Cape Canaveral and Jupiter Inlet Light.[135]

Details are recorded up to and including their arrival at City Dock near Miami, where Mrs. Henry and a friend named Miss Smith boarded on December 9, 1941. The crew sailed with them onward to Palm Beach for what appears to be the last trip any member of the Henry family would take aboard the *Odyssey*.[136]

Final entries in the log see the crew bid farewell to Mrs. Henry and Miss Smith on December 18, 1941, as they headed off again to the Merrill Stevens Boatyard, this time to begin stripping the boat throughout the remainder of the month and into early January 1942.[137]

Not long after, the Henrys and the *Odyssey* parted ways for good. While many privately owned vessels would meet the fate of being commandeered by the U.S. Navy, others were simply donated to the cause. According to William Barklie, the *Odyssey* was sold to the government in good faith for a total of one dollar.[138]

Chapter 6

WORLD WAR II, BECOMING USS *SALUDA* AND THE FIRST NAVY RESERVE RECRUITS

My dad, along with a bunch of his buddies, took boatbuilding at Miami Edison High School. They were all Depression-era kids, and sailors in Miami. Right after Pearl Harbor, it was early 1942 when he and two of them were in Coconut Grove working on a 20-foot Bahama boat when an old navy chief walked up to them and recruited them on the spot. The navy had requisitioned a 65-foot schooner named the Ashley *and they were sailing a triangle from Miami to Bimini to Fort Lauderdale, and back to Miami. The whole thing was about having eyes in the Gulf Stream looking for German submarines. One day they returned to port and everyone was abuzz about this fabulous gold-plater yacht that just showed up in town.*

—Randy Wall, son of George H. Wall[139]

On December 7, 1941, the Japanese attack on Pearl Harbor solidified the United States' involvement in World War II, and a naval fleet that began with around seven hundred commissioned vessels would rise to include more than six thousand by the war's end.[140]

With the U.S. government having requisitioned or acquired scores of privately owned pleasure and commercial vessels for wartime needs, in July 1942, the *Odyssey* was added to that number after being turned over to the navy by the Henrys.[141]

The sailboat was sent off to the Port Everglades Section Base in Fort Lauderdale, Florida, not entirely unfamiliar waters, under charge of the 7th

Naval District commandant. Here she would be outfitted to diesel auxiliary power during the remainder of the summer and officially brought into navy service that fall.[142]

Around the same time, albeit under slightly different circumstances than Barklie Henry's, George H. Wall and his two buddies Warren Servatt and Don Richardson enlisted into the navy reserves. They had all gone to Miami Edison High School together, where they earned vocational degrees in boat building. With the government desperate for candidates with sailing skills, the three young men were found by the navy about thirty miles south of Port Everglades, at Coconut Grove near Miami Beach.[143] Each was issued a uniform, completely bypassing boot camp, and was sent straight away aboard a 65-foot schooner called the *Ashley*, now owned by the navy.[144]

Part of a class of small auxiliary sailing vessels under the navy's charge, each vessel carried the identifying letters IX followed by a two-digit number between 82 and 90 on the hull. The *Ashley* was classified as IX-83. However, by 1942, most would be named after rivers or Native American tribes.[145]

According to George Wall's son, Randy, "They never went to boot camp because the *Ashley*'s mission was considered so important, and they needed guys that knew how to sail. That's not something they were going to learn at boot camp."[146]

It was not at all uncommon for the navy to recruit skilled tradesmen from civilian jobs when necessary, especially during the post–Pearl Harbor ramp-up. The branch enlisted many sailors in this manner, assigning paygrades that mirrored skill level.[147]

A number were enlisted as petty officers and chiefs right away, particularly in the areas of ship repair, radio operation and technician support.[148] It was much like the familiar Seabees, a play on the initials CB, or Construction Battalion. Formed around this time, they were largely made up of experts from the civilian sector.[149]

The navy reduced basic training to as little as three weeks right after Pearl Harbor, and the length changed twenty-eight more times during the war.[150] This makes it probable that a number of new recruits could have avoided it altogether.

Finding themselves assigned to the *Ashley*, Wall, Servatt and Richardson complemented a total crew of six. They sailed primarily from Miami to the closest part of the Bahamas, across the Gulf Stream and back again. With two guys on the cross sheets of the main and two on the foremast, there were four sets of eyes at all times looking out over the Gulf Stream for German submarine activity.[151]

The young men continued in this role until the *Odyssey* showed up at Miami's port one hot summer's day, creating quite a stir. The sleek white yacht had everyone in the section volunteering to be a part of the "gold-plater's" crew.[152] A "gold plater" was a term used to refer to yachts that had been built by Henry Nevins due to a gold-plated label that appears on his builds.[153]

"My dad and the other two fellas went to check the bulletin board and their names were there to report for duty," said Wall. "They followed up with their commanding officer and all three were immediately transferred to *Saluda*."[154]

Upon joining the navy fleet, the *Odyssey* became the USS *Saluda* IX-87 (named for a river in South Carolina). Under the command of Lieutenant Edward F. Valier, a junior commissioned officer, some members of the typically nine-man crew included George Wall and his school mates Servatt and Richardson.[155]

Beginning as seamen first class (S1c), the three were set alongside a motor machinist mate (MoMM), a cook (SC), a radioman (RM) and one more crew and officer position that would change several times over the course of the next three years.[156]

After being outfitted in Fort Lauderdale, *Saluda* was officially commissioned by the navy on June 20, 1943.[157] Rather than becoming a defensively armed naval or combatant ship, she was to be used for experimental research attached to the Bureau of Ordnance at Fort Trumbull's Underwater Sound Laboratory in New London, Connecticut.[158] As far as weapons go, she was, in fact, equipped with only one sidearm on board.[159]

The navy established an experimental station at Fort Trumbull as early as 1919, but it wasn't until 1945 that the Underwater Sound Laboratory was officially formed as a response to German U-boat threats. Over the next fifty years, it would develop remarkable innovations there in underwater sound propagation, but especially during World War II, with *Saluda* very much a part of it.[160]

While many of the newly acquired navy vessels had been used in patrolling the coast for enemy submarines, *Saluda*'s quiet wooden hull under sail, with no mechanical noise, made her virtually undetectable. She became a valuable asset for conducting sound research that would be used to improve sonar and other anti-submarine warfare technologies.[161]

This involved hosting civilian scientists from places like Woods Hole Oceanographic Institution in Massachusetts, who came aboard with all manner of devices and supplies used in sound research and testing.[162]

N. Nav. 5-b
(Jan. 1930)

Page 1

REPORT OF CHANGES

JUL 13 1943

11

of U. S. S. SALUDA

COMMISSIONING REPORT

~~for the month ending~~ 20th day of June, 19 43, date of sailing

from ______ to ______

	1 NAMES (Alphabetically arranged without regard to ratings, with surname to the left and the first name written in full)	2 SERVICE NUMBER (The service number must under no condition be omitted)	3 Rating at Date of Last Report	4 Date of Enlistment			5 Place of Enlistment
1	RICHARDSON, Donald Arthur	560-58-12	S1c	9	22	42	Miami, Fla.
2	SERVATT, Warren Thurman	560-58-13	S1c	9	22	42	Miami, Fla.
3	THOMPSON, Sam Hill	560-68-69	SC2c	11	11	42	Miami, Fla.
4	WALL, George Henry	560-58-11	S1c	9	22	42	Miami, Fla.
5	WHEATLEY, Walter Nelson	560-53-55	MoMM2c	9	1	42	Miami, Fla.

	6 Branch of service	7 Received, transferred, deserted, discharged, change of rating, death, or any other change of status	8 Date of occurrences in column 7	9 Vessel or station from which received, to what vessel or station transferred, where discharged and character of discharge; where deserted, and amount due or overpaid. Where died, cause of death and where and when buried. If rated and authority for same. If disrated, give cause; if on detached duty, give place of duty. If passenger, give purpose of travel and final disposition.
1	M-2,USNR	Rec.	6-20-43	Section Base, Port Everglades, Florida
2	M-2,USNR	Rec.	6-20-43	USS YP-534
3	M-2,USNR	Rec.	6-20-43	USS YP-534
4	M-2,USNR	Rec.	6-20-43	Section Base, Port Everglades, Florida
5	M-2,USNR	Rec.	6-20-43	Section Base, Port Everglades, Florida

Naval Muster Roll from June, 20, 1943, showing the names of Don Richardson, Warren Servatt and George Wall, Depression-era kids, boat builders and friends who were recruited by the navy for their skills. *Courtesy of the National Archives.*

On board USS *Saluda*. *Left to right*: Don Richardson, Warren Servatt, unknown and George H. Wall ("Wheels") at Woods Hole, winter of 1944. *Courtesy of Randy Wall.*

USS *Saluda*. *Left to right*: Don Richardson, Warren Servatt, unknown and George H. Wall at Woods Hole, winter of 1944. Ice on the decks. Woods Hole Oceanographic Institution (WHOI) in Massachusetts was contracted by the Bureau of Ships and the U.S. Navy Department *Courtesy of Randy Wall.*

USS *Saluda*, IX-87, from the archives of George H. Wall, shown painted gray, most likely on the East Coast, 1943. *Courtesy of Randy Wall.*

Primarily sailing and conducting experiments off the East Coast during this time, in the fall they would head to Jacksonville, Florida, to get hauled out and refit at the Mayport Navy Yard. Once that was finished, they sailed onward to the Caribbean and their base on the southern island of Tobago to avoid the New England winters and continue their work.[163]

Lieutenant William "Bill" Gallagher was an officer who came in under Valier and would eventually take over as the skipper aboard *Saluda* when Valier transferred out.[164]

The United States was well into the war by the time Gallagher joined the navy reserve in July 1942. Quickly commissioned as an ensign, the twenty-two-year-old had no experience other than "considerable amount of sailing on board his family's 46-foot schooner" written on his application.[165]

Assigned to Port Everglades, Florida, it was here that he became acquainted with Lieutenant Valier. After a brief stint aboard a 96-foot steel-hulled staysail schooner, Gallagher took command of an 84-footer called the USS *Romain* before joining *Saluda*.[166]

USS *Saluda*, IX-87, side view, painted gray and sailing. Part of a class of small auxiliary sailing vessels under the navy's charge, each carried the identifying letters IX followed by a two-digit number between 82 and 90 on their hulls. *Courtesy of SSSOdyssey.org.*

After about four months patrolling for submarines with his USS *Romain* crew between Miami, Palm Beach and the Bahamas, he met up with Valier at port. "He says to me, Bill, we're going down the Caribbean, and how would you like the go along as first lieutenant, so that's what I did," said Gallagher.[167]

In a 2012 interview, he fondly recalled three Miami boys who were a part of his crew. "There were three kids that came in the reserve as apprentice seamen. One became a quartermaster, and the other two were bosuns by the time I left. They had worked on charter boats in Miami, so they all knew boats. They were pretty good."[168] The kids he referred to were Wall, Servatt and Richardson.

A view of USS *Saluda*, IX-87, from her bow back. *Courtesy of SSSOdyssey.org.*

One time a freighter had somehow broken loose while tied up in Trinidad, proceeding to run into *Saluda*'s topside, not only bashing in a couple of planks but also breaking several frames on the outside in the process.[169] It seems the navy really had found the right boys for the job. No job was too difficult for the trained boat builders, as they set to work getting the materials needed sent to the island to make repairs. Running a navy float alongside the vessel, they were able to easily scarf in new topside planks and laminate new frames, taking care of the problem without further help or delay.[170]

IX87

Opposite: USS *Saluda*, IX-87, no sails, either motoring or at anchor. *Courtesy of SSSOdyssey.org.*

Right: Lieutenant William "Bill" Gallagher (*left*). Gallagher was an officer who came in under Valier and eventually took over as the skipper aboard *Saluda*. *Courtesy of SSSOdyssey.org.*

According to Randy Wall, the trio also got into some harmless mischief from time to time. During one of the haul outs at Mayport as *Saluda* was preparing for an upcoming trip to the Caribbean, the two officers in charge were released on leave for a few weeks.[171] Meanwhile, the others stayed behind, as civilian workers from the navy yard carried out duties like servicing the engine, bottom cleaning, going over the rigging and doing everything that needed to be done before getting underway.[172]

Left to their own devices, with no major duties of their own at the time, the trained boat builders with an innate sense and knack for fixing and building just about anything got to work on a little project.[173] "By that point my dad could requisition anything that he deemed the ship needed in the way of supplies," says Randy Wall. "So, they requisitioned a bunch of plywood, a Briggs-Stratton two-cylinder engine with a centrifugal clutch, wheels, tires, axles, and everything they needed. They proceeded to build a wooden copy of a jeep with a fold down windshield, just like a real navy jeep."[174]

George Wall pictured at the wheel of the "famous wooden jeep" (with his later wife) built by himself, Servatt and Richards. It was carried aboard USS *Saluda* and driven at port. *Courtesy of Randy Wall.*

George H. Wall stands on the bow of *Saluda*, October 28, 1945, signed by him with his nickname "Wheels." *Courtesy of Randy Wall.*

They constructed a wooden cradle to park it forward of the deck-house and aft of the mainmast and covered it in a canvas makeshift garage. "The skipper immediately wanted to know what they had put on his sailboat when he returned," says Randy amusedly. "The three boys unlaced the tarp covering, had a tackle rigged, and they picked up the main boom with a topping lift and swung this jeep ashore."[175]

Not surprisingly, they spent the rest of their time there driving the skipper all around the navy base. In fact, the makeshift jeep, which they had throughout the war, became a very useful piece of equipment at port and went with them everywhere. They drove it in Jamaica, Tobago, Miami and even Woods Hole during the summertime. Wall was now referred to as "Wheels."[176]

"They were Depression-era kids," says Randy. "Everything they ever had, they made. They made their own boats as teenagers to go sailing."[177]

According to Gallagher, by the time he officially took over as skipper there were six crew members, along with three officers and, sometimes, four or five scientists. "I had a cook. I didn't always have a radioman, but I did about half the time, and I had a motor machinist," he said.[178]

Muster roll archives indicate periodic crew changes, but a few of the names were fairly consistent, such as Wall, Servatt, Richardson, a motor machinist mate (MoMM) named Walter N. Wheatley and a cook (SC) named Sam Thompson.[179]

"I had a big fat cook from Arkansas named Sam. I liked old Sam," said Gallagher.[180] Sam Thompson shows up on *Saluda*'s muster rolls from June 1943 through February 1945. He was rated SC2, which indicates his role as ship's cook, second class petty officer.[181]

Finding themselves in some rough weather on the way to Norfolk, Sam remained in the doghouse, strapped tightly in his life vest. "I asked him, 'what's a matter, Sam?' and he said, 'you ever see those chairs stuffed with

Left: George H. Wall at *Saluda*'s bow, holding a line. It looks like winter with his heavy coat and collar pulled up. The anchor windlass (used to lower and weigh anchor) is pictured in right forefront. *Courtesy of Randy Wall.*

Right: George H. Wall shown on *Saluda*'s stern. The American flag hangs off back of the ship. You can see the mizzen mast. *Courtesy of Randy Wall.*

horsehair? I feel like I got a big wad of that in my belly,'" said Gallagher. "I think he must have thought we were going down."[182]

At night, they would split into groups and take turns standing watch, a requirement for the safety of the vessel and all on board. According to Gallagher, there were five watches—three four-hour watches and two six-hour watches. "You know, you're on from midnight to four in the morning, and they wake you up 20 minutes early, and then you're 10 minutes writing up the log. You're lucky if you get three hours sleep all night. I never drank coffee until I joined the navy, but they bring you a cup of coffee. It's pretty good just to hold it. It was nice and warm in your hands."[183]

Besides watches, the main crew of six was responsible for handling the day-to-day operations like sailing and maintenance. Even raising the ship's enormous and heavy mainsail could be a daunting task.[184]

The massive gear, made of heavy canvas back then, was hoisted using the anchor windlass. They would run the halyard through a block at the base of the deck, situated just forward of the anchor windlass, to raise the cumbersome mainsail.[185]

Although *Saluda* was used primarily as a research vessel, they often foun themselves in some interesting places with interesting assignments, and those are only the ones that are declassified. Once, as they were just underway from New London to Nova Scotia, they got a call to return to the base immediately: "It says take Secretary of the Navy to Bar Harbor, Maine," said Gallagher. "The Secretary of the Navy was James Forestall, and we sat there waiting for him, but it was just about the time the Japanese gave up, and he couldn't get away. We waited for about two weeks. He never showed up."[186]

Even though it was relatively smooth sailing for *Saluda* during most of the war years, they were still within hostile waters off the East Coast, Caribbean and Gulf of Mexico, where German U-boats had been responsible for sinking more than two hundred ships.[187]

Not surprisingly, a few close calls did occur, although no known casualties happened on *Saluda* until after the war. On July 18, 1974, off the coast of Cuba and the Bahamas at a place called Cay Sol Bank, Gallagher recounted a German U-boat encounter. It came up alongside a rocky bank to most likely to recharge the batteries that kept it propelling, while *Saluda* was anchored a few miles away.[188]

"The next morning, we had a blimp come in and challenge us, and we gave him a response," said Gallagher. "He went over, and he must have seen something. The next thing we know, he went down. It was the only blimp we lost in the whole war. That German sub dropped him."[189]

The blimp Gallagher spoke of was U.S. Navy K-Class airship *K-74*. The blimp was downed by German U-boat *U-134* after unwittingly going against orders to attack it in an effort to save nearby ships yet unawares. It was, in fact, the only blimp lost to enemy fire during World War II.[190] "The DE [destroyer escort] told us the next day to get out of there. They were looking for him [the German sub], but he was gone," said Gallagher.[191]

Outmaneuvering several attacks to thwart him, *U-134* was finally sunk on August 24 near the opening to the Bay of Biscay, a gulf in the North Atlantic off Spain.[192]

On May 8, 1945, the war finally came to an end with Germany's unconditional surrender. With the Allies victorious, V-E Day (Victory in Europe Day) celebrations began to spread. *Saluda* and the boys were at sea and making their way back from scheduled operations in the Caribbean. They would have likely known due to radio silence being lifted, if not by other means.[193]

"On the eighth of May we were getting messages all day, and not just us, but all the ships at sea," said Gallagher. "We were told if you encounter Germans, fly two black balls on the starboard yardarm, and take them to the nearest port."[194]

About a week after V-E Day, they were making their return stateside. They were about sixty miles off Cape May, New Jersey, when out of nowhere a German submarine surfaced alongside of them. "I heard the story directly from the lips of my dad when I was a kid, and the other two men later in life, who said they never expected to live through the day," said Randy Wall.[195]

In Servatt's account to Randy Wall, he was the signal officer and radioman who began intercepting the signals being sent by the German U-boat. As he used *Saluda*'s signal lamp to message back, the rest of the crew hurriedly came on deck, concerned with the situation at hand.[196]

Gallagher recalls being awakened by the quartermaster at about two o'clock in the morning. "He says to me, 'Mr. Gallagher, there's a submarine.' I figured it was probably something else, but sure enough, it was a submarine. They had challenged him, and couldn't read the signal. The submarine had maneuvered downwind of us," said Gallagher.[197]

Meanwhile, according to Servatt, as he was busy talking to the skipper and doing the signaling, he was also interpreting the signals being sent back. The message he received was a request to surrender to them.[198]

This must have seemed like a shocking request for a research vessel that wasn't really equipped to defend itself, with only the one sidearm on board, "*Saluda* wasn't just any boat though, it was a commissioned navy ship," said Randy Wall. "The skipper tells Servatt to signal back that they need to call this in to Washington first, telling his crew, we can't just do this."[199]

Servatt quickly went down below to make the radio call, receiving the following instructions to signal back to the German U-boat: "Steam directly to New York Harbor. Stand down at the entrance, and make themselves known there. A navy boat will come and escort them in."[200]

Warren recounted to Randy in later years that it was the most terrifying event that happened the entire time they were aboard between 1943 and 1945.[201]

Chapter 7

USS *SALUDA*'S ROLE IN THE DEVELOPMENT OF ANTI-SUBMARINE WARFARE

Schedule for USS Saluda *& USS* Mentor, *25 June 1946*
1130—25 June, USS Saluda *underway to anchor off Tarpaulin Cove.*
1200—25 June USS Mentor *underway to Tarpaulin Cove where calibration measurements will be made with USS* Saluda.
1630—25 June USS Saluda *to return to WHOI.*
1630—25 June USS Mentor *to proceed to 1,400 fathom curve—Atlantic Canyon area (approximately 39-30N; 70-00W) to arrive on station at about dawn on 26 June 1946; hourly stops will be made on route for the purpose of taking underwater photographs.*[202]

Saluda played an important role as a research vessel both during and after the war, and that's only from the declassified bits that have been uncovered. Data collected throughout her research would lay the foundation for early discoveries about Sound Navigation and Ranging (SONAR).[203]

With *Saluda* still attached to the U.S. Navy Underwater Sound Laboratory in New London, Connecticut, she was first used in the areas of Sound Fixing and Ranging (SOFAR) and exploring the sound channel in March 1944.[204]

Scientists Maurice Ewing and J. Worzel, out of Woods Hole Oceanographic Institution (WHOI) in Massachusetts, were contracted by the Bureau of Ships and U.S. Navy Department. Setting out aboard R/V (research vessel)

Saluda, using explosives, they tested theories about low-frequency sound and its ability to travel far at subaqueous depths in the sea.[205]

Gallagher recounted frequently hosting scientists on board and, besides being crowded, what it was like. "They were trying to get the word on sound transmission," said Gallagher. "In the very early part of the war, there were three outfits making sonar, and nobody knew why all operated on a different frequency. I had the head of the physics department at Columbia [Dr. Maurice Ewing] on board, and he thought if he got down deep enough it was practically unlimited."[206]

Woods Hole, having only come into existence in 1930, was at the beginning of establishing itself as a leader in ocean science and engineering. Much of this was due to early navy-sponsored programs like this one. The research that occurred during World War II found each of the organizations relying on the other to solve national defense problems, all while making new discoveries about the oceans.[207]

George Wall, assigned to *Saluda* throughout the war and after, talked about spending summer living at Woods Hole, located on the southwest corner of Cape Cod, with Martha's Vineyard a little farther to the south; they had a section of quarters for the crew and civilian scientists. Many would bring their wives along. Wall arranged for his own wife, whom he had married in the spring of 1944, to come and spend the summer there.[208]

Both Ewing and Worzel joined the crew aboard the ship either together or separately, assisted by other scientists for day, week or even trips occurring over multiple weeks. They set up shop in the aft cabin, or stateroom, which also served as their sleeping quarters on lengthier excursions.[209]

A number of vessels worked in tandem with *Saluda*, as a deep receiving hydrophone listening device was placed on board. A second vessel would drop explosive charges deep below the water as far as nine hundred miles away.[210]

Between March 22 and April 5, 1944, a wooden-hulled submarine chaser called *SC 665* fired charges for the *Saluda*, followed by two others, *SC 1292* and *DE 51* (destroyer escort), also known as the *Buckley*.[211]

"We got into deep water off of Nassau, Bahamas," said Gallagher. "We had a DE out of Key West and he made up a whole bunch of half ton charges. There was a pressure element, so they wouldn't detonate until they were at 4,000 ft, and every four hours they'd chuck one over the side, and we heard them all the way to Nova Scotia. In places like Barbados, there were a bunch of boats that would shove off and go out and start picking up all the dead fish," said Gallagher.[212]

George H. Wall and his wife, pictured in 1944, aboard *Saluda*, with navigation equipment. *Courtesy of Randy Wall.*

During the winter, *Saluda* and crew—including the trio of Wall, Servatt and Richardson—would head back to their base in the Caribbean and continue working from there. They often had an escort from the U.S. Navy or Coast Guard due to the significance of the classified missions.[213]

On one such occasion, they were heading south after a haul out at the Naval Station at Mayport, Florida, near Jacksonville. While the vessel enjoyed the fairest winds and seas, radioman Servatt was hailed by the escort that its skipper needed to speak to his skipper.[214]

Continuing to work nearby, he heard the two skippers over the radio: "We are at full throttle, and rolling one rail down, and then, the other. We haven't had anything to eat except cold food since we left Jacksonville. Is there any way you could slow down for a few hours so we can feed our crew?"[215]

"Warren loved telling that story," Randy Wall said. "They needed to recover from having to keep up with this sailboat that was half their length, because she was so fast."[216]

What scientists discovered from the research conducted on board the *Saluda*, and other vessels like her, was that long-range sound transmissions from a solitary explosive detonation could be heard as a series of sounds that became so sharp by the end of the transmission that even an untrained listener could hear it.[217]

This was important because while World War II was still happening, the navy was carrying out tests to equip planes and seagoing vessels with small explosive charges that would explode on impact in the vast reaches of the ocean. The sound signals had the potential to travel through the channel, being heard at listening points set up at various intervals along the coast, revealing the location for an imminent rescue.[218]

"The idea was they were going to put monitoring stations out in the Pacific, and they knew how fast sound travelled under water, it was all hush-hush," said Gallagher.[219]

The data led to pivotal research during the submarine warfare era as the navy and oceanographers conducted further studies on high-frequency acoustics and ambient noise in the development of sonar to locate mines and submarines.[220] Ewing and Worzel would go on to write books about their discoveries such as *Propagation of Sound in the Ocean*, among others.[221]

When the war finally came to an end, *Saluda* was decommissioned briefly, and the navy, as with many vessels now in surplus, offered its return to the original owners, the Henry family. Not surprisingly, the interior and deck of the ship looked a bit worse for wear by this point.[222]

"That summer, I went to Woods Hole on my mother's behalf to inspect *Odyssey* because the Navy was offering her back to us. My father had divorced by then, and remarried," said William Barklie in a letter. "It was both wonderful and shocking to stand at the dock and look down at her. Everything that had been bright and beautiful, the teak decks, the mahogany detailing, the bronze winches, everything was buried under layers of Navy paint."[223]

Besides the painted hull, there were holes in the beautiful butternut interior for hanging and placing high-tech equipment, racks where bookshelves once stood and the replacement of the original Buda engine.[224]

On his advice, William's mother left the ship with the navy. He suggested that the money needed to restore the yacht would be better spent on a smaller one that could be operated with fewer crew.[225]

Saluda, by now a familiar sight at the Navy Underwater Sound Laboratory in New London, Connecticut, returned there in August 1945. Remaining through the fall, the decommissioned ship the navy had officially inherited for good was brought back into service for the 3rd Naval District's senior commissioned officer until the following spring.[226]

It should be noted that on February 4, 1946, the only known death to have occurred on board *Saluda* was recorded by the navy. Joseph F. Kotwica, MoMM3c, "died aboard the ship in the line of duty. Cause of death—electric shock."[227]

The vessel was recommissioned in May 1946, and work commenced at WHOI, where *Saluda* was again engaged as a research vessel, this time testing and exploring the physical characteristics of the ocean, coasts and shape of the sea bed, called hydrography.[228]

Hydrographic work became important when several World War II amphibious battles shed light on the fact that the navy had outdated nautical charts and inadequate knowledge about the oceans in general. Things like local tides, depths and underwater topography could influence the success of military assaults and equipment, if used to their advantage.[229]

"We started out in Mayport and sailed through the Sargasso Sea to Saint Thomas and the Virgin Islands, Charlotte Amalie, and on to Barbados, and Port of Spain, Trinidad, the Gulf of Paria, and Venezuela's Orinoco River," said Gallagher. "We were looking for all different kinds of bottoms, mud, coral, and rock."[230]

Gallagher remembers this time as some of his best sailing. "Just before Port of Spain I jumped on a broad reach, and that was the best. I could see in that Caribbean going north and south, and that's where all those islands are. Pretty cheap transportation, nice and flat."[231]

The research they did there was important in SONAR, too, since hydrographic conditions played a role in how safely submarines maneuvered and the effectiveness of sonar.[232]

REPORT OF CHANGES

Page 1

of U. S. S. SALUDA (IX 87) (USN Underwater Sound Laboratory, Ft. Trumbull, New London, Conn.)

for the month ending 1st day of MARCH, 1946, date of sailing

from ______ to ______

	1 NAMES	2 SERVICE NUMBER	3	4	5
1	CARBINE, James F.	717 60 89	S2c		
2	KOTWICA, Joseph F.	312 06 19	MoMM3c		

	6	7	8 FEBRUARY	9
1	V6	Rec.	26	R/S, Miami, Fla.
2	USN	Death	4	Died aboard ship in the line of duty. Cause of death - electric shock. Remains sent to USNH, Key West, Florida.

M. I. LAPP,
Lieut.Comdr., USN,
Executive Officer.

Copies to:
FltRecOff., S.F., Calif.
File.

Naval Muster Roll from February 4, 1946, showing the only known death aboard *Saluda* to be recorded by the navy, that of Joseph Kotwica. *Courtesy of National Archives.*

Besides *Saluda*, the navy had other boats it loaned out as research vessels. The USS *Mentor* was another yacht it had acquired and converted for naval use, later used at WHOI.[233]

WHOI had vessels in its own fleet, too, that had been outfitted for the purpose of research. The first one was a well-known 144-foot steel-hulled ketch called *Atlantis*.[234]

On September 15, 1944, *Atlantis* and *Saluda* would become really close during the Great Atlantic Hurricane. The Category 4 storm blew up the coast while the two ships moored at the National Marine Fisheries Service Dock. They broke free, taking the dock with them.[235]

Gallagher was there and remembers *Saluda* holed up alongside *Atlantis* for protection until an abnormal tide came and pulled them out, dock and all. "There wasn't any slack in the lines," said Gallagher. "So, we broke loose. The dock broke loose, and we drifted with the wind into the only shallow area and mud flat in the bay."[236]

According to Gallagher, he gave Lambert, the captain of *Atlantis*, a jingle (on the radio). "Next thing I know, the lights in town were changing, and I said 'Lambert, we're adrift.' He said, 'we can't be.' And sure enough, we were." After *Atlantis* bumped into *Saluda*'s side, Gallagher quickly grabbed an axe and chopped all the lines.[237]

The two boats managed to land on Ram Island flats, a mud-bottomed area, where they remained for three weeks. *Atlantis* suffered only minor damage, but it cost upward of $17,000 to get her pulled out and moved where she needed to be.[238] Surprisingly, *Saluda* did not suffer any major damage besides a small area in her hull that had been stove in by the larger boat.[239]

Between May and early September 1946, records indicated that many of *Saluda*'s experiments took place off Martha's Vineyard at Tarpaulin Cove, Gay Head and a bit farther south at an undersea feature called Atlantis Canyon. This was, of course, familiar territory for *Saluda* from her previous life as the *Odyssey*.[240]

Again, working in pairs, this time it was *Saluda* and the *Mentor*. Their mission, known as Project B-16, consisted of a series of underwater photographs, calibration measurements and data collection using instruments like a thermocouple and a deep bathythermograph (BT) for taking microstructure measurements, temperature and depth.[241]

This early equipment worked by lowering the torpedo-shaped BT device from a small winch on the ship. This let out wire as it descended. Once it arrived at the deepest point, a brake would make it stop, and then it would be

Saluda and the WHOI research vessel *Atlantis* after the Great Atlantic Hurricane in September 1944. *Courtesy of SSSOdyssey.org.*

Close-up of *Atlantis* after the September 15, 1944 hurricane. The two boats had managed to land on Ram Island flats, a mud-bottomed area, where they remained for three weeks. *Courtesy of SSSOdyssey.org.*

Another view of the two vessels after the hurricane. *Atlantis* suffered only minor damage, but it cost upward of $17,000 to get her pulled out and moved where she needed to be. *Saluda* did not suffer any major damage besides a small area in her hull. *Courtesy of SSSOdyssey.org.*

Atlantis and *Saluda* in the mud flats. This image shows just how large *Atlantis* is next to *Saluda*. *Courtesy of SSSOdyssey.org.*

brought back up to the surface. The data was recorded with a temperature-detecting element and a stylus that created etch marks on a glass slide to register temperature against pressure.[242]

They learned that temperature and depth affected the direction of sound waves. During the daytime, when the surface water was warmer, sound waves actually bent downward, creating conditions that did not work as well with sonar equipment that otherwise operated fine in the morning or at night.[243]

This allowed them to make predictions, storing the data within sound profiles. They created underwater atlases of geographical regions based on the characteristics. It differentiated certain areas as safe zones that could potentially help navy submarines remain out of enemy sonar range or predict where enemies may be hiding.[244]

The combination of these discoveries significantly affected not only the outcome of the war and the future development of naval sonar systems but also the science of oceanology.[245]

PROJECT B-16

SCHEDULE for 5 June, 1946 to 8 June, 1946.

5 June.

1. 1300 USS SALUDA underway to proceed to following points:

(a) 41° 15'N - 71°11'W
(b) 39° 25'N - 71°11'W
(c) 40° 38'N - 72°19'W

(a) The scientific party will include Dr. Redfield, Mr. Bohn and Mr. Spalding with Dr. Redfield in charge.
(b) Bathythermograph lowerings will be made hourly during the triangular cruise outlined above.
(c) Mr. Bohn will perform tests with his thermocouple beginning at a depth of 100 ft. and continuing at 100-foot increments thereafter until the thermocouple's point of maximum pressure resistance has been reached. Recordings of these tests will be made by means of a power level recorder. BT lowerings will be made during each test.

2. 1630 USS MENTOR due to arrive from New London, Conn.

6 June.
USS SALUDA at sea.
USS MENTOR in port for check of experimental equipment.

7 June.
USS SALUDA at sea.
USS MENTOR in port for check at experimental equipment.

8 June.
USS SALUDA ETA Woods Hole 0800
USS MENTOR in port for check of experimental equipment.

cc to: Director W.H.O.I.
C.O. USN USLNL
Dr. Hersey
Dr. Liebermann
Dr. Redfield
Dr. Wilson
Dr. Woollard
C.O.USS SALUDA
O.M.G. USS MENTOR
Mr. Allen
Mr. Arsove
Mr. Bohn
Mr. Churchill
Mr. Spalding

F.C.Ryder
Lt.Cmdr.USNR
Project Officer

Documents from *Saluda*'s work, Project B-16, June 5–8, out of Woods Hole. This shows the names of navy personnel and scientists involved in the mission, consisting of underwater photographs, calibration measurements and data collection. *Courtesy of SSSOdyssey.org.*

After *Saluda*'s work at WHOI was finished that year, she spent the rest of the fall and winter at the Navy Underwater Sound Laboratory, soon to leave Fort Trumbull and the East Coast forever.[246]

While *Saluda* remained with the navy for the next thirty years, William "Bill" Gallagher was transferred to an APL-8, or barracks ship. Briefly

Above: William "Bill" Gallagher when he returned to visit *Odyssey* in 2012, back at the helm with the Sea Scouts in Tacoma. *Courtesy of SSSOdyssey.org.*

Left: Bill Gallagher during a visit to see *Odyssey* after many years. He was also interviewed by Michiel Hoogstede and Dick Shipley during that time, 2012. *Courtesy of SSSOdyssey.org.*

finding himself in the Philippines, by February 1946 he had left the navy reserves to head back to a life of farming, a wife and, soon, a family.[247]

With a wife and family on the way himself, George Wall mustered out just before the Korean War.[248] According to navy documents, Don Richardson and Warren Servatt were also discharged around 1946.

For *Saluda*, the next part of her journey was just getting underway. Transferred to the 11th Naval District on January 8, 1948, in San Diego, by the spring, she would undergo a major haul out at the Thames Shipyard in New London, Connecticut, in preparation for the long sail west.[249]

Chapter 8

THE SAN DIEGO YEARS

In 1948 is when they decided to ship her to San Diego because it was an officer training school, and they needed a boat so the boys could learn how to sail. The order went out to put it on a flat car and then somebody thought, railroad flatcar, and an 88-foot yawl, that's pretty damn big and high, and they could only get her as far west as someplace in New Jersey. Then the order went out to put her on the deck of a freighter and ship it around, and that order went across the desk of a reserve commander in Washington who said, "That's ridiculous. This is a boat, it ought to be sailing around, and he put together a crew.

—Don Frothingham, ensign (ENS), Navy Reserve, on Saluda*'s transit to San Diego*[250]

While WHOI and the Navy Underwater Sound Laboratory at Fort Trumbull were leaders in research on the East Coast, the Navy Electronics Laboratory and places like Scripps Institution of Oceanography in San Diego were epicenters on the West Coast, as well as *Saluda*'s next stop.[251]

Scripps, having taken the name of early founders—philanthropist Ellen Browning Scripps and brother E.W. Scripps, a newspaper mogul—was highly involved with the war effort in 1941 and a main hub of the University of California Division of War Research (UCDWR). The UCDWR supported navy defense initiatives like contracting with the nearby U.S. Navy Radio and Sound Laboratory (NRSL) for sonar research.[252]

It was established at Point Loma in 1940, and the staff was an interchangeable mix of both navy and University of California civilian scientists. Renamed the Navy Electronics Laboratory (NEL) in 1945, this would be *Saluda*'s home for the next twenty-six years.[253]

In 1948, Don Frothingham was an ensign in the Naval Reserves Officer Training Corps (NROTC). Having spent some time aboard a classic "New York 50" yacht, he became part of the crew tasked with getting *Saluda* to San Diego.[254]

According to him, the skipper's name was Commander Schufelt, who taught navigation for the navy. There were two ensigns, three lieutenants and a few more guys from the NROTC, all put up to the task for their sailing skills.[255]

"One of the lieutenants was a sharp fella named Bob Parker, Rob Varrel, and another called Wurtman, a brand-new ensign named Flint, and a NROTC named Bill Mitchell. Four were regular navy, a cook, who we just called cookie, the Bosun, Engineer, and a Radioman," said Frothingham.[256]

They met up with the ship in New London, where she was tied up across the dock from the famous, now dismasted schooner *Atlantic* on the Thames River.[257] Designed by William Gardner in 1903, the three-masted, 185-foot schooner held the world record for fastest crossing of the Atlantic under sail for many years.[258]

Before getting underway, they did a dry run to Block Island, just off Rhode Island, "We sailed on out, and it was a beautiful day. We saw a squall coming down on us, and we thought, a ship like this can take a squall like that, so we just strapped her down and sailed on in," said Frothingham. "I want to tell you, when the squall hit, of course she leaned over some. Most of the crew was used to boats coming upright. They had been on destroyers, and they always come back up. They came up on deck faster than anything I've ever seen. But they got used to it."[259]

"Cookie," being an old navy destroyer guy, set out to the New London naval base for all the provisions. "Since it was a submariner base, they got the best food of all," said Frothingham. He ordered up everything they could possibly fit on board, stuffing the rest in the lazarette. He would pick up other things along the way, going into navy installations to get whatever he wanted, "He was sharp, but he also knew how to cook. We ate very, very well," said Frothingham.[260]

On the first leg of the long journey to San Diego, their first stop was Bermuda. It was July 1948. Frothingham described it as a beautiful night with a good fresh wind coming up and a full moon as they sailed into the evening.[261]

Before too long, they were headed into another real blow, pretty common for that time of year. They ended up using a storm trysail, a

Saluda at sail in the sea and on her way to San Diego, looking rough as she heels a bit. *Courtesy of SSSOdyssey.org.*

small sail that is set to help in controlling the boat during a storm or high winds. "We were on a storm trysail so long we could have belonged to the storm trysail club. You know, sailors who have sailed their ships in a real blow," said Frothingham.[262]

There were three watches made up of three crewman each. Everyone but the cook and engineer had a shift. That night, when he took his ten o'clock watch, Frothingham found himself sailing well into the morning on his own, as the rest of the crew were bushed from earlier. "I kept an eye out for any kind of storms, and followed the moonlight path for some time. I looked at the compass, and realized I was way off, and better correct it," he said. "We landed in Bermuda in the right place, so my correction was apparently not too bad."[263]

Tying up alongside of a cruiser off Somerset in Bermuda, they took a navy commandant out for a courtesy sail. Although World War II had ended, there still was the Royal Navy Dockyard, and many ships were held over at the U.S. bases at St. David's and the Naval Air Station at Southampton.[264] After a short layover, it was on to the next stop, Puerto Rico.

For the next six days, they sailed from Bermuda to San Juan, Puerto Rico, on a port tack, the wind coming over the port side of the boat. "In the trades [trade winds] there was nothing to do but keep an eye on the steering to make sure you weren't being overtaken by a wave, but other than that, it was easy," said Frothingham.[265]

Approaching Puerto Rico, the enormous Spanish-style fort El Morro off in the distance, with almost no tide made navigating into port quite smooth. With just a few days' stopover, some of the crew took shore leave, while others remained on watch.[266]

"I remember that first night I got shore leave and a group of us went to a hotel called the Escabron," said Frothingham. "Since we had been on one tack for six days, everything sloped. Of course, we got there at 8:30 or 9:00 p.m., and nobody was around, because the party doesn't really begin until 10:30 or after." Further recollections revealed a hearty welcome by the people of the island, who were very used to seeing navy personnel by that time.[267]

From there, they sailed onward through the Caribbean Sea to Panama. It was the only part of the trip they didn't have much wind and resorted to using the motor. Reaching Colon on the Atlantic side of the Panama Canal took about seven days. *Saluda* tied up easily to the navy pier with almost no tide at all.[268] This was her second time passing from east to west through the Panama Canal.

On the other side of the canal, the Pacific side was quite the opposite. As *Saluda* was tied up to a floating dry dock, she moved up and down,

Saluda at dock, taken sometime during the journey from east to west—painted white again, with the name *Saluda* barely visible on the stern. *Courtesy of SSSOdyssey.org.*

View of *Saluda* at dock with tarp/tent to stay cool during the heat. A crew member is seen taking a rest. *Courtesy of SSSOdyssey.org.*

Saluda at dock somewhere along the long journey to San Diego. *Courtesy of SSSOdyssey.org.*

with as much as fourteen-foot tides on the way through to Balboa.[269] This phenomenon occurs because each of the oceans have varying sea levels. The Pacific side can rise up to twenty feet, whereas the Atlantic side, a mere forty-five miles away, is only three feet higher at its highest tide.[270]

As they made their way through the canal, small engines moving along tracks called "donkey lines" kept the boats straight. "The commander had to tell people running the donkey engines, 'don't yank the bits out. This is just a wooden boat, and you're used to the old freighters you can yank on all you want.'"[271]

Frothingham and some of the others ended their journey in Balboa. When all was said and done, he concluded, "There was one broken engagement, one divorce, and one postponed marriage."[272]

Most of *Saluda*'s trip north from there was under tow by a tug. "Either the navy was antsy, or their sailors were not dead game sailors, or both," said Frothingham.[273]

Finally arriving at her new home in San Diego, she was operated as a training vessel for reserve officer candidates' (ROC) seamanship instruction. In addition, she was involved in several major expeditions in the ongoing study of underwater acoustics at the Navy Electronics Laboratory (NEL).[274]

Saluda at dock, perhaps at the end of her journey west. Note the beautiful wood deck detail. *Courtesy of SSSOdyssey.org.*

In San Diego, *Saluda* was operated as a training vessel for reserve officer candidates (ROC), as well as several major expeditions in the ongoing study of underwater acoustics at the Navy Electronics Laboratory. *Courtesy of SSSOdyssey.org.*

The ROC program had rolled out in 1949 with two training stations, one in Newport, Rhode Island, and the other in San Diego. The program had two parts: a six-week summer course that began in July and an advanced course lasting another six weeks. Studying topics such as navigation, naval weapons, seamanship, communications, orientation and leadership, candidates at each location were trained to operate different types of vessels.[275]

During San Diego's summer portion, candidates boarded the carrier USS *Valley Forge* for inspection tours and cruises and learned the operation of USS *Saluda* as a part of their training.[276]

At the same time, she was intermittently being used for research. Chief Warrant Officer George Hansen was the skipper and navigator in charge from 1949 through 1952. He and a crew of five enlisted men were housed in the NEL dorms.[277] With no prior sailing experience to speak of and no training prior to the assignment on the sailboat, Hansen's sail training would be learned "on the job."[278]

Above: ROC, July 18, 1950, learning use and operations of sextant. *Courtesy of SSSOdyssey.org.*

Opposite, top: ROC personnel, July 18, 1950, receiving instruction on plotting ship's course. *Courtesy of SSSOdyssey.org.*

Opposite, bottom: ROC aboard *Saluda* learning knot tying. Notice the line in each hand. *Courtesy of the National Archives.*

ROC cadet training on raising and lowering *Saluda*'s massive sails. *SSSOdyssey.org.*

Before joining *Saluda*, though, Hansen was the harbormaster of the Pacific Island Kwajalein, located in the Marshall Islands. Prior to that, he was a quartermaster on the USS *Phoenix* during the Japanese attack on Pearl Harbor and responsible for maneuvering the cruiser out of harm's way. His experience more than made up for his lack of sailing skills at the time.[279]

Similar to *Saluda*'s World War II research, everything always seems to lead back to the connection between understanding sounds in the ocean and sonar development. NEL experiments focused on the sea bed and how underwater sound was affected by different kinds of surfaces below the water. The navy also worked with University of California and Scripps scientists to study seismic refraction.[280]

Quite a number of small ships were used for this purpose between 1949 and 1976 off San Diego's coast, in places like the Southern California Borderlands region and San Clemente Island. *Saluda* had the distinction of being the only sailboat.[281]

Carried out under well-known Scripps oceanographer Russell W. Raitt, *Saluda* was again a receiving vessel while another ship set off explosive charges. Hydrophones were set up on *Saluda*'s decks or near the surface of the water for detecting the explosions from varying distances. Much of the data collected was helpful in creating updated topographic and sonar charts for the navy, particularly for submarine use.[282]

Hansen recounted taking scientists out with their recording gear and microphones hanging over the sides of the ship. They took others out too. Mostly those connected with the lab, but sometimes important dignitaries. The trips, usually during the daytime and always during good weather, still found many guests unused to the rolling seas, becoming quite seasick.[283]

In the mid- to late 1960s, research aboard the vessel shifted to include the study of porpoises and marine mammals. Although the navy's official dolphin research programs began around 1963 at a lab in Point Miigu, California, it was realistically born much earlier.[284]

Besides bioacoustics, there were a number of reasons why the navy studied the highly intelligent and trainable mammals. By 1967, the operation had moved to Point Loma and later became the Navy's Marine Mammal Program (NMMP).[285]

In 1969, *Saluda* was part of a fourteen-day study off the coast of Mexico under the command of Chief Boatswain Russell Ludwig and a crew of eight. Research and Engineering Department staff zoologist William E. Evans led the group as they set out to study the navigational behavior of porpoises in their natural habitat.[286]

With *Saluda* as a silent platform, the team searched for porpoises to catch, tag and ultimately track after they were freed. A team of marine equipment specialists and research engineers was present to help out.[287]

About four hundred miles south of NEL, near Scammon Lagoon, they found a large population of *Delphinus bairdi*, *Delphinus delphis*, *Tursiops gilli* and *Tursiops nuuanu*, all relatives of the bottlenose dolphin.[288]

Details being taken into account during the capture period included the sex, size and shape. They attempted to collect porpoises from each of the separate herds to learn about the group's sexual composition.[289]

Making careful observations about where they were found, the scientists used instruments like a bathythermograph to gauge the depth and temperature of the water. They recorded the latitude and longitude coordinates, time of day, direction, how fast they traveled and the presence of different marine life. They were tagged with an easily identifiable plastic disc attached to the dorsal fin.[290]

Stories passed down through the years mention *Saluda* carrying a large saltwater tank on her foredeck at one time, probably in the same spot the wooden jeep was kept. A few porpoises were supposedly kept there and released on training missions carried out by NEL. Personnel would activate a radio collar worn by each so they could be tracked, although no evidence has surfaced to corroborate this.[291]

In June 1969, *Saluda* was involved in more bioacoustics studies of porpoises, whales and fish. Led by scientist Dr. William C. Cummings of the Naval Undersea Center (NUC), the goal was to distinguish sounds made by two species of whales, the finback (*Balaenoptera physalus*) and the Bryde whale (*Balaenoptera edeni*), and compare them to earlier recordings.[292]

The month-long cruise, with Chief Warrant Officer Ludwig again in command, set off from La Paz on the Baja Peninsula. Cummings was assisted by a group from NUC's listening division, research psychologists and a biology professor named Donald R. Nelson from California State College in Long Beach who studied shark acoustics.[293]

While NUC studies were conducted, Nelson recorded sounds made by fish in peril. He had successfully conducted previous experiments that lured sharks to a reef by playing noises that mimicked struggling fish. The hope was that if they could attract sharks with certain underwater acoustics, perhaps they could also deter them.[294]

From on board *Saluda*, similar experiments were attempted. Sending the frightening cries of killer whales below the surface, they tried to test whether the sounds would motivate the behavior of porpoises or other large whales. Some were as long as eighty feet, close to *Saluda*'s overall length of eighty-eight feet.[295]

Left: Lieutenant Commander Don Kidder was in charge of the Naval Undersea Center, Long Beach Division, from September 1971 through 1973, the later years of *Saluda*'s time in California. *Courtesy of Don Kidder.*

Below: Lieutenant Commander Don Kidder did not spend much time on board *Saluda*, but he was the OIC (officer in charge) of the fifteen boats there at the time—all powerboats except *Saluda*. Naval Undersea Center (NUC) is pictured on the sign in the background. *Courtesy of Don Kidder.*

May 25, 2023 photo of Don Kidder, who said "it meant so much to me to see her after all these years" when he met the author and visited *Odyssey/ Saluda* in Tacoma. *Author's collection.*

With *Saluda*'s operating equipment off, they quietly observed more than seventy-five finbacks feeding alongside hundreds of Baird's whales from deck. Using an array of hydrophones, microphones, preamplifiers and sound analyzers to isolate interfering noise, they were able to make recordings.[296]

Not far from Mulegé and Loreto, the first ever recordings of Bryde's whales were captured from *Saluda*. Two thirty-six-foot-long whales were sighted and swam away from each other when *Saluda* approached. One of them kept passing back and forth under the keel. This would be the precursor for much of Cummings's future work.[297]

Lieutenant Commander Don Kidder was in charge of the Naval Undersea Center, Long Beach Division, from September 1971 through 1973, the later years of *Saluda*'s time in California. Although he did not spend much time on board, he was the OIC (officer in charge) of the fifteen boats there at the time, all powerboats, except *Saluda*.[298]

Kidder recalled that most of the work done at Long Beach involved weapons testing: "*Saluda* was used on occasion to support our work, but we never supported their work." A few times he observed work being done with animals on trips to San Diego.[299] "San Clemente Island, where *Saluda* spent some time, was administered by NUC, and there was a whole lot going on out there that not many were privy to," says Kidder.[300]

According to him, *Saluda* was still being used in listening and monitoring submarines well into the Cold War era. One of the few times he'd gone aboard, he noted numerous batteries in the salon area of the vessel and how no effort was made to conceal them. He also saw engineering equipment, and when the boat went out, there were between six or seven Raytheon engineers on board.[301]

Besides *Saluda*'s ability to be silent, she had the perfect disguise. Not looking like a typical navy vessel and, in fact, looking like a sailboat was her greatest camouflage.

Chapter 9

YACHT RACING FOR THE NAVY AND WINNING

I sailed the Saluda *in the Newport Harbor–Ensenada 1950 International Ocean Race on 4 May 1950. We entered with an LDRR rating of 67.0, a volunteer crew of 18 Navy personnel who took leave from their jobs to make the race. At this time, the* Saluda *had only the three lower working sails, two small jib topsails, and a mizzen staysail, but a ballooner, some 30 feet short on the luff, and a half spinnaker 40 feet short on the luff were borrowed from the big ketch, Zamha. This is how we entered the race. It was conceded the Navy entry was merely a gesture of good will.*

—Lieutenant Hallie P. Rice, USN, sailing master, Saluda, *1950* [302]

Not long after *Saluda* had arrived in San Diego for duty at NEL, somebody thought it a good idea to enter the sleek yacht in the Newport to Ensenada International Race, and why not? She finally would have a chance to fulfill her destiny in the shadow of the legendary *Dorade*, as well as a somewhat illustrious career as a racing boat for the navy.

It was May 1950 when *Saluda* was entered in the 125-nautical-mile race for the first time, the third year of a race that would become an annual tradition to this day.[303]

Although Chief Warrant Officer George Hansen was in overall command of the ship, it was Lieutenant Hallie P. Rice who allegedly suggested to the 11th Naval District that *Saluda* be allowed to enter the race. It was also Rice who would lead her to victory.[304]

Saluda during dry dock in San Diego, pictured with her name on the stern. *Courtesy of SSSOdyssey.org.*

For that first race, knowing that *Saluda*'s sails were in pretty bad shape, Lieutenant Rice came around to Lieutenant Colonel Lacey Hall, who happened to live on a 94-foot ketch at the time, "I told him he was welcome to the sails," said L.C. Hall. "The one string was that I went with them. Since I was military too, if another service branch. It was a good deal all around. Thus, I became a member of *Saluda*'s racing crew for as long as I remained in San Diego."[305]

The race started from Newport Harbor midday Thursday with 133 boats. One of them was *Santana*, owned by Humphrey Bogart. At the opening of the race, the crew not only got a good look at *Santana*, another Olin Stephens design, but Bogart even stopped by *Saluda* for a pre-race

Lieutenant Colonel Lacey Hall pictured at the wheel during the 1950 Newport to Ensenada Race. *Courtesy of SSSOdyssey.org.*

drink with Hansen, which was really more of a pre-celebration drink, since both would place.[306]

"It blew like hell at the start of the 1950 Ensenada, and all that day," said Rice, with *Saluda* taking the lead about thirty minutes into the start—that is how it would stay.[307]

She made about 11.6 knots between Newport Jetty and Point Loma (about fifty-eight nautical miles), and accounts relayed that *Saluda* was ahead of the fleet only thirty minutes into the race and remained unpassed by a single yacht. This seemed a surprising feat for its crew considering the ill-fitting sails and lack of racing experience.[308]

Meanwhile, with two men manning the wheel for a tough go at steering most of the way, at times they seemed to simply drift on past the lighter boats despite their setbacks, "It was really a piece of cake, though," said Hall. "The air was loaded with guys on the ship-to-ship frequency, yakking about broken gear and cold food. We sat down at the gimbaled table and had roast beef and all the trimmings."[309]

By 6:00 a.m. the following day, *Saluda* was four and a half miles from the end of the race. Although the wind becalmed, she would end up at the finish just an hour shy of the 1948 record—but first nonetheless.[310]

By the time the second boat crossed, the crew had already got the sails stowed and cleaned *Saluda* up real nice. "We got in and quickly dropped a hook [anchor] while cleaning the yacht up," said Hall. "A whale boat came over from the race escorting Destroyer escort, and ferried us in, in relays, for hot showers. Tough life."[311]

Finishing in twenty-two hours, one minute and sixteen seconds, *Saluda* became the first navy vessel to complete an oceangoing race on the West Coast. Winning both the First to Finish Trophy and the New York Yacht Club cup for first split-rig to make it through the finish line, *Saluda* arrived with much fanfare.[312]

Along with a number of racing class honors, Bogart's yacht would come in second place for the Mayor of Ensenada Trophy (A Class).[313]

For *Saluda*, a token ceremony was held immediately following the race, with a formal presentation in December of that same year in the San Diego

Saluda's crew, "Leading the Pack, 1950." It was *Saluda*'s first time in the race, and she won. *Courtesy of SSSOdyssey.org.*

The winning crew from the 1950 Newport to Ensenada Race before they got their fancy uniforms. *Courtesy of SSSOdyssey.org.*

office of the 11th Naval District commandant Rear Admiral Wilder D. Baker. George Hansen, as officer in charge of *Saluda*, was presented with the trophy by Commodore Cliff Chapman on behalf of the Yacht Club and the Newport Ocean Sailing Association.[314]

That was it. *Saluda* and the navy had caught the racing bug. With all of the good publicity, it was decided that from then on, she would be a working vessel all week long and a racing and recreation vessel on the weekends.[315]

Don Kidder's recollection of *Saluda*'s racing days was that "it was a big deal with the brass in San Diego, and generally scorned by everyone else."[316]

People on the West Coast took notice, not only because she was fast but also because her fine lines and noteworthy design really did stand out. Soon they connected her history and pedigree as the former *Odyssey* too.[317]

The navy entered in a few overnight races around the Southern California coast, but *Saluda* didn't fare as well, however, "because we never had enough wind," said Hall. "*Saluda* could use a breeze of wind. We went fast at times, but not consistently in Southern California's generally light breezes."[318]

Above: George Hansen in 1950 receiving the trophy from Newport CA Yacht Club commodore Cliff Chapman. *Courtesy of SSSOdyssey.org.*

Right: Skipper and navigator George Hansen shows off the trophy for winning the First to Finish Trophy and the New York Yacht Club Cup for first split-rig to make it through the finish line in the 1950 Newport to Ensenada Race. *Courtesy of SSSOdyssey.org.*

George Hansen, pictured later in life, speaking at a 1998 event. *Courtesy of SSSOdyssey.org.*

After the big race, the Recreation Fund somehow managed to come up with the money for a crew uniform. Consisting of a baseball cap, navy gray-and-white T-shirt, faded blue denim pants and a matching jacket with "Saluda" printed on the back, it "made for a very smart looking yacht, as contrasted with the very rag-tag get-ups we'd worn before," said Hall.[319]

Each of the crew members was given a new knife and marlinspike covered in a leather sheath. George Hansen donated items, including a scrapbook of photos from the race and his own knife set, to the Sea Scout Ship *Odyssey* organization in the mid-2000s.[320]

Further, the navy allotted funds for new sails, including a spinnaker, genoa jib and a new main, mostly all made of Egyptian cotton—a long way from how they started. According to Rice's accounts, "These weren't just any old sails, they were Kenny Watts sails, about $8,000 worth, a large sum in those days."[321] Kenny Watts was a renowned sailmaker on the West Coast who began making sails in the late 1930s.[322]

Call it beginner's luck, but *Saluda* would go on to participate in the Newport to Ensenada race a total of twenty-five more times, but only winning that first time in 1950, "We were always the scratch boat, that told the story," said Hall. "But we had a lot of fun in all our races."[323]

Above: *From left to right*: George Hansen, Rear Admiral Davis and Lieutenant Commander Taylor at *Saluda*'s helm, showing off the new crew uniforms, 1951. *Courtesy of SSSOdyssey.org.*

Left: *Saluda* showing off her new sails during a later Newport to Ensenada Race, in which she participated twenty-five times, though only winning once. Beginner's luck? *Courtesy of SSSOdyssey.org.*

Hall was transferred from San Diego in 1951 around the time of the Korean War, saying goodbye to racing aboard *Saluda*. To his great surprise, he happened to come across the vessel around 1983 at a Tacoma Sea Scouts event, remarking, "I can't tell you how thrilled and happy I was to be aboard her again. Nostalgia, wow!"[324]

Hallie P. Rice, who was sailing master of *Saluda* during the 1950 win, was sent east for duty at the Naval Academy not long after, as stated in a letter he wrote dated May 31, 1950.[325]

CHAPTER 10

THE FINAL NAVY YEARS

Whidbey Island, Washington

Naval Undersea Center's 90-foot sailing yawl USS Saluda *(YAG-87) sailed out of San Diego Harbor last Sunday, August 18, possibly for the last time. The vessel headed north along the coast for its new anchorage off Whidbey Island near Seattle, Washington.*
—Seascope, Naval Undersea Center, Friday, August 23, 1974[326]

Saluda's racing and research days for the navy in San Diego were coming to an end. Reclassified as miscellaneous auxiliary craft YAG-87 in the late 1960s, *Saluda* was somewhat of a rarity and different than most boats in the navy fleet. "She was a white elephant. There was a lot of discussion about the cost of maintenance, and they were looking for someplace that would get more use out of her," said Don Kidder.[327]

Overhauled in September 1973 with complete rewiring, battery ventilation, hull work and the installation of a non-skid deck, by the same time the following year, she had been declared excess property. This meant that the vessel was up for redistribution just about anywhere the Department of Defense decided she was needed.[328]

That would turn out to be the Athletic and Recreation Department at Naval Air Station (NAS) Whidbey Island in Washington. Departing San Diego one month later, *Saluda* got off to a rough start. The crew, it seems, was in such a hurry to get there that they ruined the engine by running it too hard. They didn't get very far, having to be towed to San Francisco for a replacement, which was installed quickly by slicing through the roof of *Saluda*'s charthouse.[329]

Located on Puget Sound roughly thirty miles north of Seattle, Whidbey Island is the largest of the Washington State islands and at one time the largest naval air installation in the Pacific Northwest.

Plans for the base began prior to World War II as the region was in need of a defensive position for rearming and refueling navy planes. Crescent Harbor, chosen as an ideal takeoff and landing zone for its seaplanes, was established as a seaplane base. The nearby Ault Field was chosen as a site for land planes.[330]

Some of the earliest squadrons of the seaplane operation were PBY-5 and 5-A Catalinas (patrol bombers) and P5-M Marlins, all planes that could make landings on water or land. They were used as enemy surveillance planes during World War II, mostly for their long-range flight capabilities but also because they carried two thousand pounds of bombs and had machine gun mounts. Even after the war, they continued to be useful for a number of training missions.[331]

However, by the 1960s, they had been all but phased out. By the time *Saluda* arrived to Crescent Harbor near building 19, a boathouse that formerly housed support boats for the seaplane operation, the area was no longer used as a seaplane base.[332]

In 1971, NAS's Department of Morale, Welfare and Recreation (MWR) took it over, and it became known as the "Navy Marina." It kept several pleasure craft there for rental by active duty or retired navy airmen and military personnel.[333]

Cathy Hinson, a young sailing enthusiast who lived in Coupeville, was twelve years old when she recalls seeing *Saluda* for the first time. She would become a part of the Sea Explorer Ship *Whidby* under skipper Roger Sherman, moored about one hundred feet from *Saluda*. Her mother, Nancy Reynolds, who had also learned to sail, was a civilian employee at the base.[334]

Besides being used by cadets and haphazardly maintained, *Saluda* was available for rental as a part of the NAS Whidbey Island sailing program. Eligible naval and military members, their dependents and special interest groups could charter the vessel for as little as sixty dollars per day. There was a skipper and a mate placed in charge.[335]

According to stories, she was known as the "good time boat," and it started to show. With many a rowdy bunch taking her out for a bit of fun, often being inexperienced sailors, improper care resulted in damage. Running aground during low tide had caused damage to the very tip of the bow, tires that were used as fenders made way for scraping by the logs she was moored to, the

Left: Cathy Hinson was a Sea Scout on board SSS *Whidby*. Her mom worked at NAS, where she saw *Saluda* for the first time and, later, became a volunteer aboard SSS *Odyssey*, Ship 190, in Tacoma. *Courtesy of Cathy Hinson.*

Right: Hinson's mom, Nancy Reynolds, who worked as a civilian employee at NAS Whidbey Island, Washington. *Courtesy of Cathy Hinson.*

transom and boomkin had been damaged, the rail cap was in bad shape and missing in spots and some of the stanchions were bent.[336]

By 1978, the boat, building 19 and the piers used by the recreation department were all in need of major repairs. *Saluda*, at this point, had seen better days.[337]

Due to a lack of resources, building 19 continued to degrade until it was condemned in the mid-1990s and, finally, demolished. A new marina was eventually constructed in place of the old one.[338]

Saluda may have been a bit luckier than the building. The navy's funding was running low all around, and these were navy airmen, not sailors. With a minimal budget for recreational activities, they were faced with a decision: invest in the necessary repairs for *Saluda* or make improvements to the long-standing golf course.[339]

Originally built in 1949, Rocky Point Golf Course started out with six holes. Captain Bill Gallery, a naval aviator who later became the base commander, was instrumental in a number of improvements to the fairway, which was eventually renamed after him.[340]

Losing out to majestic views of Puget Sound while teeing over 150 acres of manicured green, *Saluda* was added to the surplus list.[341]

Hinson's mother was tasked by brass to locate all of the paperwork for *Saluda* so that when the time came, they would have everything ready to pass off to the next owner. She worked in the building where the files were kept and spent weeks in the basement combing through documents and piecing together whatever records she could find in the pre-computer days.[342]

Pleased by her accomplishment when she handed the commander a box full of *Saluda*-related material, he offered to take her out on the boat as a reward. After a resounding yes, she requested only to bring her daughter along.[343]

Although a stormy day kept them from taking the boat out, both Hinson and her mom got a tour before she got back to her work at the base. Hinson recounted cadets on board doing bright work and begging her mother to stay and help out. Her mother acquiescing, she was able to stay. Little did she know that after boarding *Saluda* that day, their paths would cross again. In the late 2000s, she returned as a key adult volunteer, seeing two sons and a number of youths through the *Odyssey* program.[344]

Finding a new home for the almost 90-foot *Saluda*, in need of major repairs was proving to be a difficult task for the Surplus Office. Although the office offered *Saluda* to several installations, not surprisingly, none was willing to take her on. But her luck was about to change—the navy and *Saluda* would part ways once and for all.[345]

Chapter 11

BECOMING TACOMA'S TALL SHIP

The SSS Odyssey*'s Glory Days*

I didn't learn to sail until I was fifty years old. I was taking sailing lessons at the time with my wife. We learned on a 25-foot McGregor with a sailing club in Des Moines and partly in Tacoma. I've been around boats all my life. My dad always had boats, and so did I, but not a sailboat. And so, when I was fifty, I decided I wanted to learn how to sail. When I saw Odyssey *the first time out on Puget Sound, I was thinking, "Oh, my god, I would like to sail on that." I found out who owned it, and it was the Sea Scouts, so I went down to where they were keeping it in the tide flats and introduced myself and said, "I'd like to be a part of this."*

—Bob Burns, early volunteer and Odyssey *captain for thirty years*[346]

In 1978, *Saluda* made the journey from Whidbey Island to Tacoma. There were a handful of the people who were influential in making it happen and along for the ride, many of whom are no longer able to tell the story. Besides Doug Cullen and George Leonhard, Dave Phillips, Cam Miner and several others were on board for the transit from Whidbey to Tacoma.[347]

Kent Gibson, a fourteen-year-old Sea Explorer in Cullen's crew at the time, Corsair, SES 448, has the distinction of being the only youth who was able to join the group for *Odyssey*'s transit that day.[348] "What I recall was that we met down at the Totem Marina, or City Marina. The navy sent a van to pick us up. I don't recall all of the adults that were there, but about eight to ten people. Dave Phillips, Cam Miner and a handful of council executives."[349]

Kent Gibson was a Sea Scout in Doug Cullen's SES Corsair. He was the only Sea Scout on board *Saluda* during her transit from Whidbey Island to Tacoma. *Courtesy of Kent Gibson.*

Upon their arrival at the Naval Air Station (NAS) at Whidbey Island, navy personnel brought the boat around and gave them a quick run through on how to operate it. Gibson's recollection was that no sailing was done and that they motored the whole way to Tacoma. They tied up on an end pier near Eleventh Street at what used to be Totem Marina, and the bridge had to be opened each time for the 104-foot-tall mainmast to pass through.[350]

At this point, the planned use of the ship was as a council asset. Youth from local Sea Scout ships would take turns cleaning, scrubbing and doing what work they could on the boat. The navy hadn't done much upkeep toward the end. There was a great deal of work that needed done, and all manner of equipment left behind had to be removed and holes patched.[351]

Gibson remembers the boat spending a long time at the Concrete Technology graving dock. It was pulled into a bay, and a moveable wall allowed the water to be pumped out creating a dry dock. This is how hull work and bottom paint was accomplished.[352]

Eventually, the boat was relocated to a moorage outside the bridge so the bridge did not need to be raised and lowered for every single pass through. Terry Paine's one-year plan would turn into eight, spurring what can only be called the SSS *Odyssey*'s glory days and shaping a program that would last for many years to come.[353]

Sea Scouts were a little more military-like back then, according to Gibson. "We did a lot of marching and knot tying." There was a big regatta every year where all the Sea Scout ships would compete against one another in different team building exercises. "It was very interesting and fun. It helped to build discipline and character and, for a lot of kids back then, shaped them for the rest of their lives."[354]

During the early years, after the major work had been completed, Paine enlisted some teenagers who would become the stand-in crew for *Odyssey* when guests and youth from area Sea Scout ships would come aboard.[355]

Around 1982, Jim Kaiser, sixteen years old at the time, was prompted by Paine to join. Although he never officially became a Sea Scout, he and a small group of youth around the same age would regularly crew *Odyssey* and took part in many of the activities, even some races happening around that time.[356]

Involved through 1984 and briefly around 1988, Kaiser fondly remembers looking forward to his time on the boat and even all the hard work and sanding he put into it, "It was just cool to be a part of the *Odyssey*. They are some of the best memories of my life."

Eventually, after enough youth were recruited for a program and ship that was clearly unlike any before her, *Odyssey* became its own Sea Scout ship.

Odyssey on Puget Sound with full sails and Ship 190 Sea Scouts out of Tacoma, taken in the early 1990s/mid-2000s. *Courtesy of SSSOdyssey.org.*

Sea Explorers and other Boy Scout Exploring branches had officially become coed by 1972, and by the late 1990s, they had changed the name from "Sea Explorers" to "Sea Scouts."[357]

Although Gibson would never have the chance to be a part of the *Odyssey* crew, having aged out by the time the program really got up and running, his experiences were memorable ones nonetheless. He credits Sea Scouts for instilling in him many skills, including a lifelong love of sailing.[358]

Paine continued bringing in knowledgeable volunteers to train youth on everything from navigation and sail handling to helmsmanship, ultimately, unbeknownst to them, turning them into leaders. They were the crew, and the adults were there to support, guide and supervise.[359]

Until that point, he had been the main captain, serving as lead skipper of the program. A group of men came in under his watch, each obtaining the Coast Guard–required one hundred ton license needed to operate the *Odyssey*. Some of the names associated with this timeframe and beyond include men like Cam Miner, Dick Clayton, Bob Burns, Bud Bronson, Nic Marshall, Bob Goux and Rory McDonald. Many would also serve as lead skipper at one point or another, but all would put their heart and soul into the *Odyssey*.[360]

Burns, in his fifties in the early 1980s when he first became aware of the *Odyssey*, was like so many before him, enamored of Tacoma's Tall Ship. He recalls going under the Tacoma Narrows Bridge at about ten knots when Paine gave him the helm for the first time.[361] Burns, who turned ninety-five years old in 2023, felt a noticeably similar sentiment as most who have had the chance to be a part of the *Odyssey*. It was the time of his life, and he remembers it like it was yesterday, even all of these years later.[362]

Since Boy Scouts did not have a budget for Sea Scouts, it was up to each ship to fund itself. This is the case still today. To maintain the ship and program, the *Odyssey* organization began offering cruises and charters for a nominal fee. Being a Coast Guard–inspected vessel with licensed captains and classified as a subchapter T passenger vessel enabled them to do so.

On Saturdays and Sundays, three-hour evening cruises and five-hour daytime cruises were offered to the public as guests sailed throughout the Commencement Bay area.[363]

The greatest source of revenue, however, would come from charters to Boy Scout units during the summer months. Reasonably priced weeklong charters were available to the San Juan and Canadian Gulf Islands. The Boy Scouts expected to come on board and fall in line with the crew. They participated in all phases, taking an active role in a list of duties like standing

From left to right: Bob Goux (glasses), unknown, Renée Paine, Terry Paine and Bud Bronson at an *Odyssey* event, 1998. *Courtesy of SSSOdyssey.org.*

From left to right: Bob Burns, Bob Goux, Rory McDonald, unknown and Bud Bronson during an *Odyssey* event, 1998. All Coast Guard–licensed captains on board the *Odyssey*. *Courtesy of SSSOdyssey.org.*

watch at night. It was so popular with Scout units across the United States that summers were often completely booked years in advance.[364]

Tacoma's Tall Ship was not only an experience of a lifetime, but it was also a great proponent for instilling maritime skills in youth that often led to careers in the maritime industry. In an area with vast waterborne commerce and career opportunities, the community had shown a great deal of support over the years.

The organization tried to give back as much as possible. Local charities and organizations like Kiwanis and Rotary, among others, were hosted on board. Guests could even join in with the Sea Scouts to do some of the work if they so wished.[365]

The ship took part in many of the annual events in and around the Puget Sound region. Some of these included the Daffodil Marine Parade, Tall Ships Tacoma, Seafair, Holiday Cruise on Lake Washington in Seattle, local Opening Day of Boating season festivities and the Port Townsend Wooden Boat Festival. Sea Scouts would often dress ship with flags and other decorations, give tours and even host Santa Claus.

In the summer of 1983, a crew of twelve Sea Scouts and a combination of eighteen passengers and adult volunteers took part in the Portland Maritime Festival, a momentous occasion over the Fourth of July weekend. Bob Burns was there, as were Terry and Renée Paine.[366]

Sailing from Tacoma on a voyage through open sea, *Odyssey* made its way to the Columbia River and into the Willamette, with several overnight stops made along the way there and back.[367]

"It doesn't sound like much of a trip when you talk about cars, but you're talking about boats," said Burns. "From here to go down there, you've got to go up to the Strait of Juan de Fuca, out the strait to the ocean, you've got 140 miles of ocean to travel to down to the Columbia River, and up the Columbia to the Willamette River to Portland."[368]

Besides a festival celebrating boats of all sorts, it was a meeting of Tall Ships at Willamette River Park. The *Odyssey*; the *Pride of Baltimore*, a replica of the 1800s clipper; and *Adventuress*, homeport of Port Townsend, were all present. *Odyssey* was greeted by much fanfare. The mayor showed up with roses and gifts, according to Renée Paine's account.[369]

It had been a while since her racing days with the navy in California, but *Odyssey* would go on to participate in several races under the command of the Sea Scouts. In 1984, 1985 and 1994, she took part in the Swiftsure International Yacht Race, a prestigious long-distance Pacific Northwest sailboat race beginning and ending in Victoria, British Columbia.[370]

In 1984, while coming back from the race full sails, her spinnaker front and center, a Canadian Coast Guard boat in awe of her beauty came alongside. "They offered to take one of our crew members out away from the boat so they could get pictures of it," said Bob Burns, who was on board at the time. "So that's what Rory [McDonald] did, and it's one of the best photos ever taken of *Odyssey* [seen on the cover of this book]." Although she didn't win, Burns recalled that she did place pretty well.[371]

The following year, not quite making it to the Swiftsure light due to lack of wind, Burns explains, "The wind just died off. The boats with mylar sails, which were really light sails, could go on and finish, but our sails were so big and heavy, and not for light winds, so we didn't finish that particular race."[372]

The 1994 race would end even less successfully. Treacherous swells, currents, a bout with a whirlpool and weather had the ship rolling so steep that the boom hit the water several times. This resulted in *Odyssey* dropping out of the race somewhere near Tatoosh Island and Cape Flattery for the safety of crew and ship.[373]

In 1986, *Odyssey* and her crew went on another kind of adventure. Having spent a great deal of time sailing in Canadian waters during summer long cruises, she was invited to participate in a thirty-day circumnavigation of

Another stunning full-sail shot of *Odyssey* on the beautiful waters of Puget Sound with SSS 190, *Odyssey* Sea Scouts. *Courtesy of SSSOdyssey.org.*

Vancouver Island in celebration of the Canadian Expo 86. This trip, too, would take them into open ocean, sailing on Vancouver Island's west side.[374]

A group of adults and seventeen Sea Scouts cruised around the island with about fifty boats, making stops and filling up the various harbors to enjoy music and festivities.[375]

"Most of the boats were Canadian," according to Bob Burns. "There were a few American boats, and we were the biggest one, of course." Remaining with the same group of 350 people, docking in certain places, sometimes the weather would keep them over night, since many of the smaller boats would need to wait out the storms.[376]

According to Renée Paine, while traveling around the top of Vancouver Island very early one morning, a lighthouse keeper came on the radio and said of *Odyssey*, "I just want you to know that is the most beautiful sight I have ever seen."[377]

In 1988, *Odyssey* marked its fifty-year anniversary. The Sea Scouts celebrated a few years later with the return of the original carved panels depicting Homer's *Odyssey*, the very carvings Mr. Henry had gotten made for his yacht, the ship's namesake. Removed before handing the ship over to the navy at the onset of World War II, they had remained with the family all this time.[378]

Bob Burns was in his fifties in the early 1980s and just learning to sail when he first became aware of the *Odyssey*, later becoming one of its captains for the next thirty years. Today ninety-five years old, he is pictured here with the author. *Author's collection..*

Linda Barnwell, daughter of Barklie Henry, returned the original panels to the ship around *Odyssey*'s fifty-year anniversary, where they have remained since. *Courtesy of Rory McDonald.*

William Barklie Henry had made his home in California for many years, crossing paths with the *Odyssey* during his adventures north, as well as in and around Washington's coastal waters. He established a connection with some of *Odyssey*'s volunteers. Burns remembers speaking to him on the phone about the carvings, which William Barklie felt should be returned to their rightful place on board. After talking it over with family members, he arranged for his daughter Linda Barnwell, of Portland, to return the carvings.[379]

Delivered to Burns's home in Tacoma, an understanding was determined: the carvings would remain on board as long as the Sea Scouts owned the ship.

A ceremony took place at the Port of Tacoma shortly after to commemorate the occasion. Although William Barklie was not in attendance, he and Burns exchanged several letters over the years. Linda and a few other family members who lived locally were able to come. The carvings remain in *Odyssey*'s main salon today, where they were perfectly carved out to fit.[380]

The SSS *Odyssey* program would hold steady at its homeport of Tacoma. Most people knew her by now and could recognize her from a distance. The really lucky ones had stood below her 104-foot mainmast, amid billowing sails, if even for an hour or two.

Continuing to grow and introduce youth to once-in-a-lifetime experiences on the water each year, more captains and volunteers would come—each surely deserving of their own chapter for the indelible impact they, too, made on the program and on young adults from around the region.

The Mount Rainier Council eventually came to be known as Pacific Harbors Council. Around 1991, Tom Rogers, who took over SSS 110, the *Charles N. Curtis*, after Paine had left for the *Odyssey*, brought the Tacoma Youth Marine Center to fruition. Inviting *Odyssey* to the newly established dock space across the Thea Foss Waterway for a generally modest rate, she has moored there ever since.[381]

Chapter 12

THE *ODYSSEY* TODAY

Tacoma's Tall Ship, a Waterfront Fixture Since 1978, Now at Risk of Being Deep-Sixed

A 90-foot wooden sailboat that has trained aspiring Tacoma mariners on Puget Sound for more than four decades is in peril of being retired if money can't be raised to install a new mainmast.

—Peter Talbot, News Tribune[382]

Sea Scout Ship *Odyssey* would continue to operate in much the same manner for many years. However, in March 2021, a series of unfortunate events left the historic ship becalmed in more ways than one.

It was routine maintenance of her 104-foot Sitka Spruce mainmast that exposed dry rot, setting imminent removal in motion. That, combined with a worldwide pandemic bringing everything to an abrupt halt shortly after, put an end to charters of any kind, as well as the program's main income source.

With *Odyssey* shut down until it was safe to resume training, it was months before Sea Scouts could return to their regular weekend meetups. Even then, no mainmast meant no sailing, which was quite a blow for a youth sailing outfit.

With no money coming in to help with the immense cost of replacing it, the future of Tacoma's Tall Ship was beginning to look very uncertain.

Yet even in the face of adversity, something amazing happened. Although she was no longer officially a Tall Ship without her mainmast, new youth continued to join. SSS 190 still trained on board every weekend as they

The last Sea Scout Ship *Odyssey* crew, 2022, pictured on board in Tacoma, Washington. The author's son is in the back row to right of the youth at the helm. *Author's collection.*

always had, motoring around the sound and often further for weekends and activities. The faithful youth crew still participated in local events, gave tours and shared the *Odyssey* legacy.

At the end of 2021, several alumni parents, and those of current youth crew members, joined together to form a nonprofit called the Friends of *Odyssey*. Their mission is to support regional youth through the art of Tall Ship sailing as a fundraising and donation platform for the *Odyssey* program and training vessel. Not an easy undertaking, especially with their first task of trying to raise enough money for a replacement mast that shipwrights estimated at around $150,000.

Fast-forward eight short months later, and a massive community-wide fundraising campaign was launched. With a large pledge received from a private donor, the organization was given a short window of time to match the donation in order to secure the funds.

Reaching out to news media sources, several helped to spread the word, and some of the crew members were even interviewed and featured on the local TV news.

Right: The author's son, Max Molina, at *Odyssey*'s helm—one of many youths to be inspired by the *Odyssey* program over a span of forty-five years. *Author's collection.*

Below: *Front row, left to right*: Emily Molina, Captain Eric Kiesel and a volunteer. *Second row*: *Odyssey*'s youth crew, including Max Molina (second from left). *Author's collection.*

By the end of the summer, with approximately one week left to raise the necessary proceeds, the organization, through the community's generous support, did the impossible. It raised more than $150,000, quite an accomplishment for a nonprofit in existence for less than a year. It was a happy time. The group and youth were more hopeful than they had been in a long while. It had been a roller coaster, and they seemed to be on the downward slope.

With a semiannual dry dock already set for December 2022, research began on another quandary: locating the right wood for the right price to build a new mainmast.

From the very beginning, a number of problems came out of the woodwork, so to speak. From finding the necessary wood, shipping the wood and finding a company willing to sell the wood, it was looking like there would be no wood available in time for the anticipated dry dock.

Once upon a time, dry dock had taken place right in Tacoma. Recent years had seen facilities either close down or lack large enough equipment to lift the heavy vessel out of the drink. *Odyssey* would have to transit to Port Townsend.

Still trying to secure the wood and even contemplating various other options such as changing the material it was constructed out of (which would create a new set of problems), dry dock moved forward. The outfit accepted the fact that a mainmast would likely have to wait for installation at a later date.

Sometime between the ship arriving at Port Townsend Boat Haven and the Coast Guard beginning its in-depth inspections, a firm lead on wood came to light. The timing, however, would not line up with what was to come.

It had been a few years since the last dry dock, and this time around, things were not going very smoothly. No one could have predicted a number of costly fixes that would become necessary right off the bat in order for the ship to pass the Coast Guard inspection. The timetable of only a few weeks in dry dock soon turned into months.

More concerning were major repairs needed to the transom and horn timber due to areas of dry rot—work that would have to be completed before a new mainmast could even be fitted. Although the Coast Guard would allow a deadline of the following year for completion of the additional work, by this point, costs were becoming well over what had been budgeted.

It was 2023 by now, and leaders of the organization and nonprofit had to assess the best course of action and make a difficult decision. Should they

Odyssey after a very long dry dock in Port Townsend, 2022, and the discovery of much-needed restoration projects. Pictured here much like she looks today, without her 104-foot mainmast. *Author's collection.*

Odyssey moored on the Foss Waterway in Tacoma, Washington, near the iconic Murray Morgan Bridge, 2023. *Author's collection.*

Drone footage taken by Darren Chromey on the Thea Foss Waterway in Tacoma during one of *Odyssey*'s farewell charters in 2023, with beautiful Mount Rainier in the background. *Courtesy of Darren Chromey.*

continue to invest funds the Sea Scouts did not have into an eighty-five-year-old wooden ship that would likely need continued repairs? Or pay for what was necessary and required by the Coast Guard to legally get the vessel back to Tacoma and consider *Odyssey*'s retirement?

The second option was the most sensible, but not because it was what anyone truly wanted. It had been a long couple of years full of many ups and downs and not much sailing. It was the only feasible solution to keep the sailing program alive and get youth sailing again.

By the middle of March, with the ship finally back in Tacoma, the decision was made to retire *Odyssey* from Sea Scouting after more than four decades serving youth in the Puget Sound region. It was time to consider finding a replacement vessel.

That summer, farewell cruises were scheduled for the community, alumni and anyone formerly connected to the once grand yacht for a last cruise and a sad goodbye.

It was yet unknown what was to become of Tacoma's once Tall Ship.

AFTERWORD

They say that wooden boats will only break your heart, and I know this to be true. They also fill your life with some of the most challenging, exhilarating and timeless experiences, memories and people.

I began writing this book at what really looked like the end for *Odyssey*. It was a difficult and often heartbreaking position to be in. Here I was, starting my research at the beginning of *Odyssey*'s life, chronicling her journey through the annals of history as her life in Tacoma as a Sea Scout ship was coming to an end. I was living in the here and now, seeing her circumstances in real time.

There were many days I wanted to give up because it was too sad. It was sad to think of the Sea Scout Ship 190 program without the *Odyssey*, sad to think of Tacoma without its once Tall Ship and saddest of all not knowing what might become of her.

However, throughout that journey, one thing I was continually reminded of was the number of people with a connection to this old wooden boat. It fueled the fire within me. I knew that preserving *Odyssey*'s rich history was more important now than ever and would remain long after the ship was gone.

Many paths would cross mine. Many stories would be shared as it became clearer and clearer just how many lives had been touched by this one sailboat. I have always maintained that although the book is about the *Odyssey*, it's really about the people.

Emily Molina at the helm with Captain Michiel Hoogstede on Puget Sound, aboard SSS *Odyssey*, Ship 190, Tacoma, Washington. *Author's collection.*

So, when the very difficult decision was made to retire our once grand lady, the bigger problem became what to do with her. It was a recurring dilemma throughout her life. First with the Henry family and then with the navy on more than one occasion, and now it was our turn in Tacoma.

It always came back to the knowledge that she simply had too much history to end up in a scrapyard somewhere. The real question was: who could afford such a massive restoration project?

There were interested parties, of course, each aware of the great expense involved. There was even more talk—lots and lots of talk. Our lady needed action, though, and a great deal of TLC that we couldn't afford to give to her.

In the spring of 2023, an interesting turn of events was spurred by a former Sea Scout, now a shipwright with connections to that very same East Coast firm from her inception. Having gotten wind of the group's intention to sell *Odyssey*, plans were set in motion for the firm to travel west and look at the boat.

Spending the day in Tacoma hosted by members of the organization, the outfit was very impressed with what they saw. The condition of the vessel, given her age, was outstanding. The ship, its pedigree and valuable history would align nicely with a project the firm was trying to get underway back in Newport, Rhode Island.

If ever there was a silver lining, this might have been it. People were more hopeful than they had been for quite some time, and it looked like a way to truly honor the *Odyssey* legacy, returning her to her roots for what could be a whole new beginning. No one could think of a finer ending to *Odyssey*'s story.

But as the clock ticked, it began to look like another dead end, with the fate of *Odyssey* hanging in the balance. Even as I write this to you, our dear reader, today and henceforward I urge you to seek the answer to one question alone after reading what has been written here in these pages about the extraordinary journey of Tacoma's Tall Ship: "Whatever happened to the *Odyssey*?"

And remember, even when she's gone, we'll still have the memories.

Author's Notes

William "Bill" Gallagher died in 2020 but did visit the *Odyssey* (*Saluda*) one last time in Tacoma, Washington, around 2012 when the interviews used in this book took place courtesy of Michiel Hoogstede and Richard Shipley.

Losing touch with Don Richardson, George Wall passed away around 1975 when his son Randy was just twenty-one years old. Warren Servatt, who lived well into his nineties, continued to stay in touch, sharing the stories of their time together during World War II aboard USS *Saluda*.

Randy Wall, who today lives in Florida, made the trek west in the mid-2000s to visit the historic ship he'd heard so much about from his father and his comrades. He wanted to see where his dad had slept in the forecastle bunk all those nights aboard the sailboat during the war. The author was connected with him by Van Chestnutt and conducted several interviews via telephone.

Don Frothingham, ensign in the NROTC, passed away on September 1, 2022. He made one last visit to USS *Saluda* in 2009 in Tacoma, where he was interviewed by Michiel Hoogstede. He remarked how emotional it was for him to see her once again, rechristened the *Odyssey* once more.

Don Kidder, former lieutenant commander of the Naval Undersea Center, Long Beach Division, who lives in Eastern Washington today, visited the *Odyssey* in Tacoma in May 2023. During that time, the author interviewed him about his time with *Saluda*. He, too remarked how much it meant to him to come aboard her after all of these years.

NOTES

Chapter 1

1. Bailey, "Terry Paine."
2. Naval Undersea Center, "Navy Declared Excess Form."
3. Molina, interview 1 with Renée and Terry Paine.
4. Ibid.
5. Ibid.
6. Johnson, "Sea Scouts Snare Super Sailboat."
7. Molina, interview 1 with Renée and Terry Paine.
8. Ibid.
9. Ibid.
10. Ibid.
11. Ibid.
12. Molina, interview 2 with Renée and Terry Paine.
13. "Odyssey History," document from early *Odyssey* organization, n.d.
14. Ibid.
15. Molina, interview 1 with Renée and Terry Paine.
16. Johnson, "New Beginning Due Scouts' Sailboat."
17. Ibid.
18. Ibid.
19. Johnson, "Scouts Launch Nearly Free Odyssey."
20. Molina, interview 1 with Renée and Terry Paine.
21. Ibid.
22. Johnson, "New Beginning Due Scouts' Sailboat."

Chapter 2

23. Henry, Letter to the SSS *Odyssey* Crew, 1998.
24. *New York Times*, "Yachts Dot Island Harbors."
25. Tsuchiya, "First Defense of the America's Cup."
26. Reynolds, "New York Yacht Club Station 10," 15–25.
27. *New York Times*, "Barbara Whitney Introduced at Ball."
28. *New York Times*, "Miss Whitney Wed to Barklie Henry."
29. Encyclopedia.com, "Women in World History."
30. Ibid.
31. Vanderbilt Cup Races, "Profile: Harry Payne Whitney."
32. *New York Times*, "H.P. Whitney Left $62,808,829 Estate."
33. *Time*, "Harvard Drubbed."
34. *New York Times*, "W. Barklie Henry, Philadelphian, Dies."
35. Brown, "Corinthian," 39–40.
36. Encyclopedia.com, "Women in World History."
37. *New York Times*, "Miss Whitney Wed to Barklie Henry."
38. National Endowment for the Humanities, "Evening Star."
39. Back Bay Houses, "241 Beacon."
40. Henry, *Deceit*.
41. Norris, "Doll House."
42. Back Bay Houses, "241 Beacon."
43. Hemingway et al., *Letters of Ernest Hemingway*, vol. 3, Roster of Correspondents.
44. Hemingway et al., "To Barklie McKee Henry, 14 July 1927," 3:256–57.
45. Massachusetts Historical Society—Ellery Sedgewick Papers, "Collection Guides."
46. Hemingway et al., "To Barklie McKee Henry, 9–13 August 1927," 3:262–63.
47. Dignity Memorial, "William Henry Obituary."
48. *Time*, "Directors: Good Worker."
49. L., "When the Barklie McKee Henry Estate Was for Sale."
50. Ibid.
51. Ibid.
52. Ibid.

Chapter 3

53. *New York Times*, "New Yawl Is Launched."
54. *New York Times*, "New Yachts Are Named I."
55. Henry, Letter to the SSS *Odyssey* Crew, 1998.
56. Ibid.
57. Abbott, "Sparkman & Stephens Yachts."
58. Mystic Seaport Museum Collections & Research, "Roderick Stephens Collection."
59. Dorade, "History."
60. Letter to Mr. and Mrs. Paine, "Thanks So Much for Your Warm Letter."
61. *New York Times*, "Henry Nevins Dies."
62. *Time*, "Modern Living."
63. Lloyd, "Roderick Stephens, 85, Sailor and Innovator."
64. Henry, Letter to the SSS *Odyssey* Crew, 1998.
65. Ibid.
66. Ibid.
67. Ibid.
68. Mystic Seaport Museum Collections & Research, "Henry B. Nevins, Inc. Shipyard Collection."
69. Ibid., "Roderick Stephens Collection."
70. Navsource, "Auxiliary Minesweeper Photo Archive."
71. *New York Times*, "New Yawl Is Launched."
72. *Yachting Magazine*, "Odyssey" (November 1938).
73. Ibid.
74. Henry, Letter to the SSS *Odyssey* Crew, 1998.
75. *Yachting Magazine*, "Odyssey" (November 1938).
76. Ibid.
77. Ibid.
78. Ibid.
79. Ibid.
80. Henry, Letter to the SSS *Odyssey* Crew, 1998.
81. Rogers, "Nevins Yacht Yard."

Chapter 4

82. Whitney, in *Those Early Years*, 8–9.
83. Henry, Letter to the SSS *Odyssey* Crew, 1998.

84. Trumbull, "Odyssey Makes Port in Miami."
85. Henry, Letter to the SSS *Odyssey* Crew, 1998.
86. Ibid.
87. Cromer, "Great Fleet of Sailing Yachts," 5.
88. Henry, Letter to the SSS *Odyssey* Crew, 1998.
89. Henry, interview, 1986.
90. Ibid.
91. Henry, Letter to the SSS *Odyssey* Crew, 1998.
92. Henry, interview, 1986.
93. Hemingway Home & Museum, "His Life."
94. Henry, Letter to the SSS *Odyssey* Crew, 1998.
95. Hemingway Home & Museum, "Our Garden & Grounds."
96. Henry, Letter to the SSS *Odyssey* Crew, 1998.
97. Hemingway Home & Museum, "The Pilar."
98. Henry, Letter to the SSS *Odyssey* Crew, 1998.
99. Ibid.
100. Ibid.
101. Ibid.
102. Ibid.

Chapter 5

103. Henry, Letter to the SSS *Odyssey* Crew, 1998.
104. Trumbull, "Odyssey Makes Port in Miami."
105. Ibid.
106. Henry, Letter to the SSS *Odyssey* Crew, 1998.
107. Naval History and Heritage Command, "Unofficial Navy Certificates."
108. Henry, Letter to the SSS *Odyssey* Crew, 1998.
109. Henry, interview, 1986.
110. Trumbull, "Odyssey Makes Port in Miami."
111. Henry, Letter to the SSS *Odyssey* Crew, 1998.
112. Ibid.
113. Ibid.
114. Ibid.
115. Ibid.
116. Ibid.
117. Trumbull, "Odyssey Makes Port in Miami."
118. Henry, interview, 1986.

119. Trumbull, "Odyssey Makes Port in Miami."
120. Ibid.
121. Galapagos Islands, "Post Office Bay."
122. Ibid.
123. Trumbull, "Odyssey Makes Port in Miami."
124. Ibid.
125. Ibid.
126. Ibid.
127. Henry, Letter to the SSS *Odyssey* Crew, 1998.
128. McIntyre, *Odyssey* log book, August 21, 1941.
129. Ibid.
130. Ibid.
131. Henry, Letter to the SSS *Odyssey* Crew, 1998.
132. Garas, "Citizen Sailors."
133. Henry, Letter to the SSS *Odyssey* Crew, 1998.
134. McIntyre, *Odyssey* log book, August 21, 1941.
135. Ibid.
136. Ibid.
137. Ibid.
138. Henry, Letter to the SSS *Odyssey* Crew, 1998.

Chapter 6

139. Molina, interview with Randy Wall, 2022.
140. Martin, "Pluck, Pogy, and Portland."
141. Naval History and Heritage Command, "Saluda."
142. Ibid.
143. Molina, interview with Randy Wall, 2022.
144. Ibid.
145. U.S. Navy Auxiliary Ships and French Warships, "USS Small IX: Auxiliary Schooners."
146. Molina, interview with Randy Wall, 2022.
147. Faram, "Midway, Momentum and Manpower."
148. Ibid.
149. Ibid.
150. Ibid.
151. Molina, interview with Randy Wall, 2022.
152. Ibid.

153. New York 32, "Nevins-Draft 2020."
154. Molina, interview with Randy Wall, 2022.
155. National Archives and Records Administration, "Saluda (IX-87)."
156. Ibid.
157. Naval History and Heritage Command, "Saluda."
158. Ibid.
159. Molina, interview with Randy Wall, 2022.
160. Clark, "Sound Lab at Fort Trumbull."
161. Molina, interview with Randy Wall, 2022.
162. Ibid.
163. Ibid.
164. Hoogstede and Shipley, interview with William Gallagher.
165. Ibid.
166. Ibid.
167. Ibid.
168. Ibid.
169. Molina, interview with Randy Wall, 2022.
170. Ibid.
171. Ibid.
172. Ibid.
173. Ibid.
174. Ibid.
175. Ibid.
176. Ibid.
177. Ibid.
178. Hoogstede and Shipley, interview with William Gallagher.
179. National Archives and Records Administration, "Saluda (IX-87)."
180. Hoogstede and Shipley, interview with William Gallagher.
181. National Archives and Records Administration, "Saluda (IX-87)."
182. Hoogstede and Shipley, interview with William Gallagher.
183. Ibid.
184. Molina, interview with Randy Wall, 2022.
185. Ibid.
186. Hoogstede and Shipley, interview with William Gallagher.
187. New England Historical Society, "U-Boat Attacks."
188. Hoogstede and Shipley, interview with William Gallagher.
189. Ibid.
190. Wiberg, "U-134 Under Hans-Günther Brosin."
191. Hoogstede and Shipley, interview with William Gallagher.

192. Wiberg, "U-134 Under Hans-Günther Brosin."
193. Molina, interview with Randy Wall, 2022.
194. Hoogstede and Shipley, interview with William Gallagher.
195. Molina, interview with Randy Wall, 2022.
196. Ibid.
197. Hoogstede and Shipley, interview with William Gallagher.
198. Molina, interview with Randy Wall, 2022.
199. Ibid.
200. Ibid.
201. Ibid.

Chapter 7

202. Ryder, *Schedule for USS* Saluda *& USS* Mentor.
203. Discovery of Sound in the Sea, "History of the SOFAR Channel."
204. Ibid.
205. Ewing and Worzel, *Long Range Sound Transmission*, 1–76.
206. Hoogstede and Shipley, interview with William Gallagher.
207. Woods Hole Oceanographic Institution, "History & Legacy."
208. Molina, interview with Randy Wall, 2022.
209. Ibid.
210. Discovery of Sound in the Sea, "History of the SOFAR Channel."
211. Ryder, *Schedule for USS* Saluda *&* SC 665, SC 1292, DE 51.
212. Hoogstede and Shipley, interview with William Gallagher.
213. Molina, interview with Randy Wall, 2023.
214. Ibid.
215. Ibid.
216. Molina, interview with Randy Wall, 2022.
217. Discovery of Sound in the Sea, "History of the SOFAR Channel."
218. Ibid.
219. Hoogstede and Shipley, interview with William Gallagher.
220. Discovery of Sound in the Sea, "History of the SOFAR Channel."
221. Ewing and Worzel, *Long Range Sound Transmission*, 1–76.
222. Henry, Letter to the SSS *Odyssey* Crew, 1998.
223. Ibid.
224. Ibid.
225. Ibid.
226. Naval History and Heritage Command, "Saluda."

227. National Archives and Records Administration, "Saluda (IX-87) Muster Rolls."
228. Naval History and Heritage Command, "Saluda."
229. Musemeche, "Mary Sears' Pioneering Ocean Research."
230. Hoogstede and Shipley, interview with William Gallagher.
231. Ibid.
232. U.S. Geological Survey, "USGS Bridging Generations."
233. NavSource, "Mentor (PYc 37)."
234. WHOI Data Library and Archives, "Atlantis (1931 to 1964)."
235. Woods Hole Oceanographic Institution, "Great Atlantic Hurricane of 1944."
236. Hoogstede and Shipley, interview with William Gallagher.
237. Ibid.
238. Woods Hole Oceanographic Institution, "Great Atlantic Hurricane of 1944."
239. Hoogstede and Shipley, interview with William Gallagher.
240. Ryder, *Schedule for USS* Saluda *& USS* Mentor.
241. Ibid.
242. Scripps Institution of Oceanography, "Ocean in Motion."
243. U.S. Geological Survey, "USGS Bridging Generations."
244. Ibid.
245. Clark, "Sound Lab at Fort Trumbull."
246. Naval History and Heritage Command, "Saluda."
247. Hoogstede and Shipley, interview with William Gallagher.
248. Molina, interview with Randy Wall, 2022.
249. Naval History and Heritage Command, "Saluda."

Chapter 8

250. Hoogstede, interview with Don Frothingham.
251. Scripps Institution of Oceanography, "Scripps History."
252. U.S. Department of the Navy, "Fifty Years of Research and Development."
253. Ibid.
254. Hoogstede, interview with Don Frothingham.
255. Ibid.
256. Ibid.
257. Ibid.

258. Atlantic Schooner, "History."
259. Hoogstede, interview with Don Frothingham.
260. Ibid.
261. Ibid.
262. Ibid.
263. Ibid.
264. *Royal Gazette*, "Documenting the History of the Naval Annex."
265. Hoogstede, interview with Don Frothingham.
266. Ibid.
267. Ibid.
268. Ibid.
269. Ibid.
270. Panama Canal, "Challenge of Connecting Two Oceans of Different Levels."
271. Hoogstede, interview with Don Frothingham.
272. Ibid.
273. Hall, Letter to Mr. Terry Paine.
274. Naval History and Heritage Command, "80-G-418711 USS Saluda (IX-87)."
275. Bureau of Naval Personnel Information Bulletin, "All Hands, October 1949."
276. Ibid.
277. Clayton, "George Hansen and the Saluda."
278. Ibid.
279. Ibid.
280. Shor, Raitt and McGowan, "Seismic Refraction Studies."
281. Ibid.
282. Shor, essay in *Scripps Institution of Oceanography*.
283. Clayton, "George Hansen and the Saluda."
284. U.S. Naval Institute, "Oceans: Sleek Sailors."
285. Ibid.
286. Shields, "Centerites Set Sail on Porpoise Hunt."
287. Ibid.
288. Ibid.
289. Ibid.
290. Ibid.
291. Glein, "Odyssey, Tacoma's Historic Explorer Ship," 58–61.
292. Cummings, "Saluda Monitors Porpoise, Whale Communications."
293. Ibid.

294. Ibid.
295. Ibid.
296. Cummings, Thompson and Ha, "Sounds from Bryde, and B. Physalus."
297. Cummings, "Saluda Monitors Porpoise, Whale Communications."
298. Kidder, interview by Emily Molina.
299. Ibid.
300. Ibid.
301. Ibid.

Chapter 9

302. Rice, Letter to Sparkman & Stephens.
303. Hall, Letter to Mr. Terry Paine.
304. Ibid.
305. Ibid.
306. Clayton, "George Hansen and the Saluda."
307. Hall, Letter to Mr. Terry Paine.
308. Ibid.
309. Ibid.
310. Ibid.
311. Ibid.
312. Wood, "Newport to Ensenada International Race," 65–66.
313. Ibid.
314. "U.S. Navy Yawl Saluda Winner of 1950 Newport-Ensenada Race," n.d.
315. Hall, Letter to Mr. Terry Paine.
316. Kidder, interview by Emily Molina.
317. Rice, Letter to Sparkman & Stephens.
318. Hall, Letter to Mr. Terry Paine.
319. Ibid.
320. Clayton, "George Hansen and the Saluda."
321. Hall, Letter to Mr. Terry Paine.
322. *Los Angeles Times*, "Southland Sailing."
323. Hall, Letter to Mr. Terry Paine.
324. Ibid.
325. Rice, Letter to Sparkman & Stephens.

Chapter 10

326. *Seascope*, "Saluda Sets Sail for Whidbey Island."
327. Kidder, interview by Emily Molina.
328. Naval Undersea Center, "Navy Declared Excess Form."
329. "Odyssey History," document from early *Odyssey* organization, n.d.
330. NAS Whidbey Island History, "Installations."
331. National Park Service, "Historic America Buildings Survey—Naval Air Station Whidbey Island."
332. Ibid.
333. Ibid.
334. Molina, interview with Cathy Hinson, 2023.
335. *Whidbey News-Times*, "Navy Sells Its Boats at Marina."
336. "Odyssey History," document from early *Odyssey* organization, n.d.
337. Molina, interview 1 with Renée and Terry Paine.
338. National Park Service, "Historic America Buildings Survey—Naval Air Station Whidbey Island."
339. "Odyssey History," document from early *Odyssey* organization, n.d.
340. Waller, "Golf Course Was a Labor of Love."
341. "Odyssey History," document from early *Odyssey* organization, n.d.
342. Molina, interview with Cathy Hinson, 2023.
343. Ibid.
344. Ibid.
345. "Odyssey History," document from early *Odyssey* organization, n.d.

Chapter 11

346. Molina, interview with Bob Burns, 2022.
347. Molina, interview with Kent Gibson, 2023.
348. Ibid.
349. Ibid.
350. Ibid.
351. Ibid.
352. Ibid.
353. Molina, interview 2 with Renée and Terry Paine.
354. Molina, interview with Kent Gibson, 2023.
355. Molina, interview with Jim Kaiser, 2023.
356. Ibid.

357. Sea Scouts BSA, "History."
358. Molina, interview with Kent Gibson, 2023.
359. Kim, "Not Just a Cruise."
360. Molina, interview with Bob Burns, 2022.
361. Ibid.
362. Ibid.
363. Kim, "Not Just a Cruise."
364. Ibid.
365. Bailey, "Terry Paine."
366. Johnson, "Voyage Around to Portland."
367. "One Day Under the Mast," 1983.
368. Molina, interview with Bob Burns, 2022.
369. Johnson, "Voyage Around to Portland."
370. Clark, "Sea Explorers Tackle Swiftsure Race."
371. Molina, interview with Bob Burns, 2023.
372. Ibid.
373. Clark, "Sea Explorers Tackle Swiftsure Race."
374. Molina, interview with Bob Burns, 2022.
375. Molina, interview 2 with Renée and Terry Paine.
376. Molina, interview with Bob Burns, 2022.
377. Molina, interview 2 with Renée and Terry Paine.
378. Bailey, "Odyssey Returns to Fine Condition."
379. Molina, interview with Bob Burns, 2022.
380. Ibid.
381. Molina, interview 2 with Renée and Terry Paine.

Chapter 12

382. Talbot, "Tacoma's Tall Ship."

BIBLIOGRAPHY

Abbott, Brendan. "Sparkman & Stephens Yachts Combine Age-Old Features with New Technology." European CEO, June 28, 2015. https://www.europeanceo.com/business-and-management/sparkman-stephens-yachts-combine-age-old-features-with-new-technology.

"Article about Oren McIntyre, Maine Sailor Who Worked for the Henrys." Unknown clipping, n.d.

Atlantic Schooner. "History." https://www.schooner-atlantic.com/atlantic-history.html.

Back Bay Houses. "241 Beacon." October 16, 2021. https://backbayhouses.org/241-beacon.

Bailey, John. "Odyssey Returns to Fine Condition." *News Tribune*, May 7, 1990.

———. "Terry Paine: A Good Scout." *News Tribune*, n.d.

Brown, Harry. "Corinthian." In *The History of American Yachts and Yachtsmen*. New York: Spirit of the Times Publishing Company, 1901.

Bureau of Naval Personnel Information Bulletin. "All Hands, October 1949." April 10, 2019. https://media.defense.gov/2019/Apr/10/2002112369/-1/-1/1/AH194910.pdf.

Clark, Cathy Ann. "The Sound Lab at Fort Trumbull, New London, CT, 1945–1996." Scitation, Acoustical Society of America ASA, May 18, 2015. https://asa.scitation.org/doi/10.1121/2.0000172.

Clark, Tim. "Sea Explorers Tackle Swiftsure Race." *Nor'Westing: The Pacific Northwest Yachting Magazine* (July 1994).

Clayton, Dick. "George Hansen and the Saluda." *Due North* newsletter, n.d.
———. "SSS Odyssey." *Due North* newsletter, 2004.
Cromer, Charles. "Great Fleet of Sailing Yachts Being Tuned Here for Nassau Races." *Miami Herald*, 1938.
Cummings, William C. "Saluda Monitors Porpoise, Whale Communications." *Seascope, Naval Undersea Research and Development Center*, no. 27 (July 11, 1969).
Cummings, William C., Paul O. Thompson and Richard Cook. "Underwater Sounds of Migrating Gray Whales, Eschrichtius Glaucus (COPE)." Scitation, Acoustical Society of America ASA, November 1, 1968. https://asa.scitation.org/doi/10.1121/1.1911259.
Cummings, William C., Paul O. Thompson and Samuel J. Ha. "Sounds from Bryde, and B. Physalus, Whales in the Gulf of California." NMFS Scientific Publications Office. https://spo.nmfs.noaa.gov/sites/default/files/pdf-content/1986/842/cummings.pdf.
Dignity Memorial. "William Henry Obituary—Pacific Grove, CA." https://www.dignitymemorial.com/obituaries/pacific-grove-ca/william-henry-6183609.
Discovery of Sound in the Sea. "History of the SOFAR Channel." February 11, 2022. https://dosits.org/science/movement/sofar-channel/history-of-the-sofar-channel.
———. "World War II: 1941–1945." June 16, 2021. https://dosits.org/people-and-sound/history-of-underwater-acoustics/world-war-ii-1941-1945.
Dorade. "History." December 6, 2022. https://dorade.org/history.
Due North. Newsletter, 2003.
Encyclopedia.com. "Women in World History." November 28, 2022. https://www.encyclopedia.com/women/encyclopedias-almanacs-transcripts-and-maps/whitney-gertrude-vanderbilt-1875-1942.
Ewing, Maurice, and J. Lamar Worzel. *Long Range Sound Transmission*. Woods Hole, MA, 1944.
Faram, Mark D. "Midway, Momentum and Manpower—The Navy's Bureau of Personnel in World War II." Naval History and Heritage Command, U.S. Navy. https://www.navy.mil/Press-Office/News-Stories/Article/2645303/midway-momentum-and-manpower-the-navys-bureau-of-personnel-in-world-war-ii.
Galapagos Islands. "Post Office Bay—Floreana Island." https://www.galapagosislands.com/floreana/post-office-bay.html.

Garas, Daniel. "Citizen Sailors: A History of the U.S. Navy Reserve." DVIDS, April 11, 2019. https://www.dvidshub.net/news/357850/citizen-sailors-history-us-navy-reserve.
Glein, Linda. "The Odyssey, Tacoma's Historic Explorer Ship Provides Lessons for All Ages." *Nor'Westing: The Pacific Northwest Yachting Magazine* (April 1994).
Hall, L.C., USAF (Ret.). Letter to Mr. Terry Paine. "The Saluda." Pasadena, CA, November 2, 1983.
Hasty Pudding Institute of 1770. "Alumni." https://www.hastypudding.org/alumni.
Hemingway Home & Museum. "His Life." https://www.hemingwayhome.com/his-life.
———. "Our Garden & Grounds." https://www.hemingwayhome.com/our-architecture.
———. "The Pilar." https://www.hemingwayhome.com/lore.
Hemingway, Ernest, Sandra Whipple Spanier, Robert W. Trogdon, Albert J. DeFazio, Miriam B. Mandel, Kenneth B. Panda, Rena Sanderson, J. Gerald Kennedy and Rodger L. Tarr. "To Barklie McKee Henry, 14 July 1927." In *The Letters of Ernest Hemingway: 1926–1929*. Vol. 3. Cambridge, UK: Cambridge University Press, 2016, 256–57.
———. "To Barklie McKee Henry, 9–13 August 1927." In *The Letters of Ernest Hemingway: 1926–1929*. Vol. 3. Cambridge, UK: Cambridge University Press, 2016, 262–63.
Henderson, Amy. "What the Great Gatsby Got Right About the Jazz Age." Smithsonian Institution, May 10, 2013. https://www.smithsonianmag.com/smithsonian-institution/what-the-great-gatsby-got-right-about-the-jazz-age-57645443/#:~:text=of%20High%20Flapperdom%3F-,F.,Tales%20of%20the%20Jazz%20Age.
Henry, Barklie McKee. *Deceit: A Novel*. Boston: Small, Maynard & Company, 1924.
Henry, William Barklie. Letter to the SSS *Odyssey* Crew. "Recollections of My Experiences on the Odyssey." October 1998.
———. Personal interview, transcribed document. Berkley, California, December 1986.
History.com. "The Roaring Twenties." April 14, 2010. https://www.history.com/topics/roaring-twenties/roaring-twenties-history.
Hoogstede, Michiel. Interview with Don Frothingham, USS *Saluda*, USNR, sailed USS *Saluda* on transit from New London, Connecticut, to Balboa, Panama. Personal, 2009.

Hoogstede, Michiel, and Richard Shipley. Interview with William Gallagher, USS *Saluda*, Navy Skipper. Personal, May 10, 2012.

Johnson, Bruce. "New Beginning Due Scouts' Sailboat." *News Tribune*, April 23, 1981, sec. Business/Maritime.

———. "Scouts Launch Nearly Free Odyssey." *News Tribune*, April 28, 1980.

———. "Sea Scouts Snare Super Sailboat." *News Tribune*, December 12, 1978.

———. "Voyage Around to Portland Will Be a New 'Odyssey.'" *News Tribune*, June 24, 1983.

Kidder, Don. E-mail to Emily Molina. "Recollections of Saluda, Naval Undersea Center." June 6, 2023.

———. E-mail to Emily Molina. "Recollections of Saluda, Naval Undersea Center." May 29, 2023.

Kim, Wiseman M. "Not Just a Cruise…It's an Odyssey." *Valley Daily News*, n.d.

L., Zach. "When the Barklie McKee Henry Estate Was for Sale." Old Long Island, November 30, 2009. http://www.oldlongisland.com/2009/11/when-barklie-mckee-henry-estate-was-for.html.

Letter to Mr. and Mrs. Paine. "Thanks So Much for Your Warm Letter." Home of William Barklie Henry. Berkley, CA, March 18, 1987.

Lloyd, Barbara. "Roderick Stephens, 85, Sailor and Innovator in Yacht Design." *New York Times*, January 12, 1995. https://www.nytimes.com/1995/01/12/obituaries/roderick-stephens-85-sailor-and-innovator-in-yacht-design.html.

Los Angeles Times. "Southland Sailing: New Trophy Will Honor Late Sailmaker Watts." October 17, 1986. https://www.latimes.com/archives/la-xpm-1986-10-17-sp-5459-story.html.

Lozada, Carlos. "The Economics of World War I." NBER, January 1, 2005. https://www.nber.org/digest/jan05/economics-world-war-i.

Martin, Kali. "Pluck, Pogy, and Portland: Naming Navy Ships in World War II: The National WWII Museum: New Orleans." October 25, 2020. https://www.nationalww2museum.org/war/articles/naming-navy-ships-in-world-war-ii.

Massachusetts Historical Society—Ellery Sedgewick Papers. "Collection Guides." *Atlantic Monthly* (July 1989). https://www.masshist.org/collection-guides/view/fa0008.

McIntyre, Oren. *Odyssey* log book, August 21, 1941.

Molina, Emily. Personal interview 1 with Renée and Terry Paine, former *Odyssey* Sea Scout skipper and volunteer. November 28, 2022.

———. Personal interview 2 with Renée and Terry Paine, former *Odyssey* Sea Scout skipper and volunteer. November 28, 2022.

———. Personal interview with Bob Burns, *Odyssey* captain and volunteer for thirty years. November 21, 2022.
———. Personal interview with Bob Burns, *Odyssey* captain and volunteer for thirty years. October 26, 2023.
———. Personal interview with Bud Bronson, naval architect, former skipper, captain and *Odyssey* committee member. October 24, 2022.
———. Personal interview with Cathy Hinson. August 16, 2023.
———. Personal interview with Don Kidder, lieutenant commander, Naval Under Sea Center, Long Beach Division, 1971–73. May 25, 2023.
———. Personal interview with Jim Kaiser, early youth crew and *Odyssey* volunteer. July 7, 2023.
———. Personal interview with Kent Gibson, former Sea Scout, SSS 448 Corsair. June 14, 2023.
———. Personal interview with Randy Wall, son of George H. Wall, USS *Saluda*, naval recruit. February 16, 2023.
———. Personal interview with Randy Wall, son of George H. Wall, USS *Saluda*, naval recruit. October 20, 2022.
Musemeche, Catherine. "Mary Sears' Pioneering Ocean Research Saved Countless Lives in WWII." Smithsonian Institution, 2022. https://www.smithsonianmag.com/history/mary-sears-pioneering-ocean-research-saved-countless-lives-wwii-180980325.
Mystic Seaport Museum Collections & Research. "Henry B. Nevins, Inc. Shipyard Collection." May 30, 2016. https://research.mysticseaport.org/coll/spcoll028/#head61801040.
———. "Roderick Stephens Collection." June 15, 2020. https://research.mysticseaport.org/coll/coll163/#:~:text=Biography%20of%20Roderick%20Stephens&text=(1909%2D1995)%20was%20a,off%20Cape%20Cod%20in%201919.
NAS Whidbey Island History. "Installations." https://cnrnw.cnic.navy.mil/Installations/NAS-Whidbey-Island/About/History/#:~:text=On%20Sept.,Whidbey%20Island%20was%20duly%20commissioned.
National Archives and Records Administration. "Saluda (IX-87), 10/16/42-4/15/43." https://catalog.archives.gov/id/192185753?objectPage=128.
———. "USS Saluda (IX-87) Muster Rolls 6/20/43-6/1/46." https://catalog.archives.gov/id/192185753.
National Endowment for the Humanities. "Evening Star. [Volume] (Washington, D.C.) 1854–1972, March 27, 1925, Page 9, Image 9."

News About Chronicling America RSS. N.p.: W.D. Wallach & Hope, March 27, 1925. https://chroniclingamerica.loc.gov/lccn/sn83045462/1925-03-27/ed-1/seq-9/#date1=1777&index=16&rows=20&words=Barklie+Henry&searchType=basic&sequence=0&state=&date2=1963&proxtext=Barklie+Henry+&y=0&x=0&dateFilterType=yearRange&page=1.

National Park Service. "Historic America Buildings Survey—Naval Air Station Whidbey Island, Boat House." https://memory.loc.gov/master/pnp/habshaer/wa/wa0600/wa0628/data/wa0628data.pdf.

Naval History and Heritage Command. "80-G-418711 USS Saluda (IX-87)." https://www.history.navy.mil/our-collections/photography/numerical-list-of-images/nhhc-series/nh-series/80-G-418000/80-G-418711.html.

———. "Saluda." September 2, 2016. https://www.history.navy.mil/research/histories/ship-histories/danfs/s/saluda.html.

———. "Unofficial Navy Certificates." https://www.history.navy.mil/browse-by-topic/heritage/customs-and-traditions0/unofficial-navy-certificates.html.

Naval Undersea Center. "Navy Declared Excess Form for U.S.S. Saluda." July 9, 1974.

NavSource. "Auxiliary Minesweeper Photo Archive." http://www.navsource.org/archives/11/19idx.htm#:~:text=Records%20show%20that%20YMS'%20were,in%20a%20typhoon%20off%20Okinawa.

———. "Mentor (PYc 37)." http://www.navsource.org/archives/12/1437.htm.

Nemy, Enid. "A Whitney Who Shuns Glamour for a Life of Quiet Satisfaction." *New York Times*, June 30, 1974. https://www.nytimes.com/1974/06/30/archives/a-whitney-who-shuns-glamour-for-a-life-of-quiet-satisfaction-a.html.

New England Historical Society. "U-Boat Attacks of World War II: 6 Months of Secret Terror in the Atlantic." January 12, 2023. https://newenglandhistoricalsociety.com/u-boat-attacks-of-world-war-ii-6-months-of-secret-terror-in-the-atlantic.

New York 32. "Nevins-Draft 2020." http://newyork32.org/wp-content/uploads/2020/04/nevins-history.pdf.

New York Times. "Barbara Whitney Engaged to Marry; Betrothed to Barklie Mckee Henry of Philadelphia, a Senior in Harvard." July 18, 1923. https://nyti.ms/3gJYljF.

———. "Barbara Whitney Introduced at Ball; 1,000 Greet Debutante Daughter of Mr. and Mrs. Harry P. Whitney at Fifth Av. Home.

ELABORATE Floral Display Most of Tho Season's Buds Among the Array of New York Society in Huge Ballroom." January 5, 1922. https://nyti.ms/3AVqB9S.

———. "H.P. Whitney Left $62,808,829 Estate; Owned $64,155,422 Securities—Value of Stocks Now Lower—That of Bonds Higher. State Tax Is $9,513,013 Appraisal Here Omits Breeding Farms and Race Horses Owned in Other States." July 20, 1934. https://nyti.ms/3OWTdoT.

———. "Henry Nevins Dies; Yacht Builder, 72; Designer Also of Naval Vessels Constructed Yawls for Many Leaders in the Sport." January 7, 1950. https://www.nytimes.com/1950/01/07/archives/henry-nevins-dies-yacht-builder-72-designer-also-of-naval-vessels.html.

———. "Miss Whitney Wed to Barklie Henry; Notable Gathering of Persons in Society Attend Ceremony in Roslyn, L.I." June 26, 1924. https://nyti.ms/3UhiY4j.

———. "New Yachts Are Named I; Loomis's 12-Meter to Be Known as Northern Light." April 2, 1938. https://www.nytimes.com/1938/04/02/archives/new-yachts-are-named-i-loomiss-12meter-to-be-known-as-northern.html?searchResultPosition=1.

———. "New Yawl Is Launched; Mrs. Henry's Yacht, Christened Odyssey, Goes into Water." October 23, 1938. https://www.nytimes.com/1938/10/23/archives/new-yawl-is-launched-mrs-henrys-yacht-christened-odyssey-goes-into.html?searchResultPosition=1.

———. "W. Barklie Henry, Philadelphian, Dies; Retired Banker Stricken at Palm Beach—Family Socially Prominent. Father Was a Bar Leader His Only Son Is Husband of Daughter of the Late Harry Payne Whitney." December 26, 1930. https://nyti.ms/3gGmdEV.

———. "Yachts Dot Island Harbors as Sport Rises in Popularity; Large Number of Big Craft Anchored at Glen Cove Includes Morgan's Corsair—New Unit of Sailboats on Waters at Seawanhaka Many Yachts Fill Harbors of Island E.F. McCanns Plan a Cruise New Sail Boat Class." July 18, 1937. https://timesmachine.nytimes.com/timesmachine/1937/07/18/429454611.html?pageNumber=54.

Norris, Stephen. "The Doll House: Wealth and Women in the Gilded Age." Journeys into the Past, Miami University, September 8, 2017. https://sites.miamioh.edu/hst-journeys/2017/09/the-doll-house-wealth-and-women-in-the-gilded-age.

"Odyssey History." Document from early *Odyssey* organization, n.d.

"One Day Under the Mast." Unknown clipping, 1983.

Panama Canal. "Challenge of Connecting Two Oceans of Different Levels." https://www.yourpanama.com/panama-canal.html.
Reynolds, Richard J. "New York Yacht Club Station 10." *Long Island Forum: The Magazine of Long Island's History and Heritage* 60, no. 3 (1997).
Rice, Hallie P., Lieutenant, USN. Letter to Sparkman & Stephens. "Newport to Ensenada Race." Coronado, CA, May 31, 1950.
Rogers, Debbie. "The Nevins Yacht Yard—Builder of the NY 32 Class in 1936." NewYork32, 2020. http://newyork32.org/wp-content/uploads/2020/04/nevins-history.pdf.
Royal Gazette. "Documenting the History of the Naval Annex." September 13, 2011. https://www.royalgazette.com/archive/lifestyle/article/20110913/documenting-the-history-of-the-naval-annex/#:~:text=The%20NAS%20Bermuda%20Annex%20was,from%20the%20mainland%20in%20Southampton.
Ryder, F.C., Lieutenant Commander, USNR, Project Officer. *Schedule for USS* Saluda *&* SC 665, SC 1292, DE 51. Woods Hole, MA, 1944.
———. *Schedule for USS* Saluda *& USS* Mentor. Woods Hole, MA, 1946.
Scripps Institution of Oceanography. "The Ocean in Motion: Studies in Physical Oceanography." https://publishing.cdlib.org/ucpressebooks/view?docId=kt109nc2cj&chunk.id=d2_1_ch11&toc.depth=100&toc.id=ch11&brand=eschol.
———. "Scripps History." https://scripps.ucsd.edu/about/history.
Seascope, Naval Undersea Research and Development Center. "Saluda Sets Sail for Whidbey Island" (August 23, 1974).
Sea Scouts BSA. "History." December 21, 2018. https://seascout.org/about/history/#:~:text=1998%2D%20The%20Boy%20Scouts%20of,and%20was%20renamed%20Sea%20Scouts.
Seattle Times. "Donald Frothingham | Obituary." https://obituaries.seattletimes.com/obituary/donald-frothingham-1086652103.
Shields, Jim. "Centerites Set Sail on Porpoise Hunt." *Seascope, Naval Undersea Research and Development Center* (March 14, 1969).
Shor, Elizabeth Noble. Essay. In *Scripps Institution of Oceanography: Probing the Oceans, 1936–1976*. San Diego, CA: Tofua Press, 1978.
Shor, George G., Russell W. Raitt and Delpha D. McGowan. "Seismic Refraction Studies in the Southern California Borderland, 1949–1974." eScholarship. San Diego: University of California, April 27, 2010. https://escholarship.org/uc/item/8h49r3cm.
Talbot, Peter. "Tacoma's Tall Ship, a Waterfront Fixture Since 1978, Now at Risk of Being Deep-Sixed." *News Tribune*, August 16, 2022. https://www.thenewstribune.com/news/local/article264562506.html.

Time. "Directors: Good Worker" (January 30, 1939). https://content.time.com/time/subscriber/article/0,33009,760724,00.html.

———. "Harvard Drubbed" (June 30, 1924). https://content.time.com/time/subscriber/article/0,33009,718646,00.html.

———. "Modern Living: As Idle as a Painted Ship" (July 12, 1954). https://content.time.com/time/subscriber/article/0,33009,861010-1,00.html.

Trumbull, Stephen. "Odyssey Makes Port in Miami with Fish Eyes for Hospital." *Miami Herald*, May 2, 1940.

Tsuchiya, Steven. "The First Defense of the America's Cup." *Scuttlebutt Sailing News*, August 7, 2020. https://www.sailingscuttlebutt.com/2020/08/06/the-first-defense-of-the-americas-cup.

U.S. Department of the Navy. "Fifty Years of Research and Development on Point Loma, 1940–1990." Google Books. Naval Ocean Systems Center. https://books.google.com/books/about/Fifty_Years_of_Research_and_Development.html?id=ppjfAAAAMAAJ.

U.S. Geological Survey. "USGS Bridging Generations with WWII Technology." https://www.usgs.gov/news/featured-story/usgs-bridging-generations-wwii-technology.

U.S. Naval Institute. "Oceans: Sleek Sailors—The Navy's Marine Mammal Program." February 21, 2019. https://www.usni.org/magazines/proceedings/2007/may/oceans-sleek-sailors-navys-marine-mammal-program.

U.S. Navy Auxiliary Ships and French Warships. "USS Small IX: Auxiliary Schooners." https://www.shipscribe.com/usnaux/IX3/IX082.html.

U.S. Navy Electronics Laboratory Newsletter. 1950. Private collection.

Vanderbilt Cup Races. "Profile: Harry Payne Whitney." April 24, 2019. https://www.vanderbiltcupraces.com/blog/article/vanderbiltcupraces.com_profile_harry_payne_whitney.

Waller, Jim. "Golf Course Was a Labor of Love from One Navy Captain." *Whidbey News-Times*, April 1, 2016. https://www.whidbeynewstimes.com/crosswind/golf-course-was-a-labor-of-love-from-one-navy-captain.

Whidbey News-Times. "Navy Sells Its Boats at Marina." March 3, 2005. https://www.whidbeynewstimes.com/news/navy-sells-its-boats-at-marina.

Whitney, Conner Gertrude Vanderbilt. Essay. In *Those Early Years*. New York: Turtle Point Press, 1999.

WHOI Data Library and Archives. "Atlantis (1931 to 1964)." https://www.dla.whoi.edu/ships/776503d6-babd-4478-a209-f886d9b13bd2.

Wiberg, Eric. "U-134 Under Hans-Günther Brosin July 1942 Bahamas Patrol." Eric Wiberg—Nautical Author and Historian, March 7, 2017. https://ericwiberg.com/2014/04/u-134-under-hans-gunther-brosin-july-1942-bahamas-patrol.

Wilma, David. "Oak Harbor—Thumbnail History." HistoryLink.org, July 30, 2007. https://www.historylink.org/File/8223.

Wood, Don. "Newport to Ensenada International Race." *Yachting* (June 1950).

Woods Hole Oceanographic Institution. "The Great Atlantic Hurricane of 1944." September 5, 2010. https://www.whoi.edu/multimedia/the-great-atlantic-hurricane-of-1944.

———. "History & Legacy." January 17, 2019. https://www.whoi.edu/who-we-are/about-us/history-legacy.

———. "John I. Ewing." November 16, 2001. https://www.whoi.edu/who-we-are/about-us/people/obituary/john-i-ewing.

Yachting Magazine. "Odyssey" (November 1938).

ABOUT THE AUTHOR

Courtesy Heidi Simanjuntak Photography.

Emily Molina is a freelance writer in the South Puget Sound region of Washington, where she has lived with her retired army veteran husband and two children since 2014. She is a former international flight attendant now specializing in lifestyle, food and travel writing, and her work has been published in numerous media outlets including *South Sound* magazine, *Northwest Travel & Life* magazine, *Northwest Yachting* magazine, *48 Degrees North* magazine and *425* magazine, among others. A lover of both history and the sea, with a thirst for learning about people and those who came before, she is often drawn to historic places, like the *Odyssey*, in this, her first book. It was chasing a story that led her to the *Odyssey*, creating a new love and magnificent memories and inspiring her to learn how to sail.